NEW TRENDS IN TISSUE ENGINEERING AND REGENERATIVE MEDICINE - OFFICIAL BOOK OF THE JAPANESE SOCIETY FOR REGENERATIVE MEDICINE

Edited by **Hideharu Hibi** and **Minoru Ueda**

New Trends in Tissue Engineering and Regenerative Medicine - Official Book of the Japanese Society for Regenerative Medicine
http://dx.doi.org/10.5772/58896
Edited by Hideharu Hibi and Minoru Ueda

Contributors

Hideharu Hibi, Takanori Iwata, Tomoyuki Kawase, Hideto Saijo, Katsu Takahashi, Misako Nakashima, Akihito Yamamoto, Yoshinori Sumita, Wataru Katagiri, Masamitsu Oshima, Takashi Tsuji

Published by InTech

Janeza Trdine 9, 51000 Rijeka, Croatia

Publishing Process Manager Danijela Duric

Technical Editor InTech DTP team

Cover InTech Design team

First published September, 2014

Printed in Croatia

Additional hard copies can be obtained from orders@intechopen.com

New Trends in Tissue Engineering and Regenerative Medicine - Official Book of the Japanese Society for Regenerative Medicine, Edited by Hideharu Hibi and Minoru Ueda

p. cm.

ISBN 978-953-51-1724-7

INTECH

open science | open minds

Contents

Preface

This book has been edited through collection of all the achievement performed by the main members of The Forum of Tissue Engineering in Dentistry which was established in 2000 and it brought together the specific experiences of the scientific community as well as the clinical experiences of the most renowned experts in the field of all over dentistry. The editors are especially proud of bringing together the leading biologists and dentists of all specialties. Taken together, this unique collection of world-wide expert achievement and experiences represents the current spectrum of possibilities in tissue engineered substitution in not only dentistry and also medicine.

Historically, tissue engineering, which aims at regenerating new tissues, as well as substituting lost organs by making use of autogenic or allogeneic cells in combination with biomaterials, is an emerging biomedical engineering field. There are several driving forces that presently make tissue engineering very challenging and important: 1) the limitations in biological functions of current artificial tissues and organs made from man-made materials alone, 2) the shortage of donor tissues and organs for organ transplantation, 3) recent remarkable advances in regeneration mechanisms made by molecular biologists, as well as 4) achievements in modern biotechnology for large-scale tissue culture and growth factor production.

The idea of tissue engineering is not quite as new as it seems. The Nobel Laureate Alexis Carrel performed seminal work in the early 1900s that paved the way for today's tissue engineers. Carrel even caught the imagination of the pilot Charles Lindbergh. After his historic first solo flight across the Atlantic Ocean, Lindbergh worked with Carrel at the Rockefeller Institute in New York, with the goal of maintaining viable tissue and organs *in vitro* for subsequent implantation *in vivo.*

The earliest attempts at engineering tissue were carried out in skin by Bell, Yannas, and Green at the Massachusetts Institute of Technology in the late 1970s, early 1980s.Dr. Iannas Yannas, collaborated in studies both in the laboratory and in humans to generate a tissue-engineered skin substitute using a collagen matrix to support the growth of dermal fibroblasts. Dr. Eugine Bell later transferred sheets of keratinocytes with fibroblasts referring to them as contracted collagen gels. From a more clinical aspect, one of the most exciting advances in culturing skin epithelium was carried out by Dr. Howard Green and it has been the ability to use these cells in treating patients with burns and other skin disorders. The technique of culturing human skin epithelium in defined media and the exciting possibility to multiply epithelial cells up to ten thousand fold the amount of the original skin sample has been available for more than three decades by now. Many scientific pioneers in the field of biology, surgery and other disciplines have added their efforts with enormous creativity and ambition to further improve this method. The introduction of this technique into burn treatment in the early eighties of the past century was followed by initial enthusiasm about the prospect of saving the lives of many extensively burned patients. Our early success in Nagoya experiences have started by following the Green's artificial skin. Because he kindly provided the 3T3-J2 cell line and it supported our research works.

Tissue engineering was catapulted to the forefront of public awareness with the airing of a BBC broadcast on the potential of tissue-engineered cartilage using images of the now-infamous "mouse with the human ear", fondly referred to as auriculosaurus, from the laboratory of Dr. Charles Vacanti at University of Massachusetts Medical Center. This has become known as tissue engineering, which were all based on the same premise, that new functional replacement tissue could be generated from living cells seeded onto appropriately configured scaffoldings. In the example of cartilage, viable chondrocytes were seeded onto porous polymer fibers and configured in the shape of the desired tissue. Other potential applications of tissue engineering include the replacement of worn and poorly functioning tissues as exemplified by replacement of small caliber arteries, veins, bone and cartilage; replacement of the bladder, muscle and nerve tube; and restoration of cells to produce necessary enzymes, hormones, and other bioactive secretory products, such as salivary gland.

In spite of significant scientific progress in tissue engineering, there are few examples of human application. Two potential explanations for this may be 1) problems associated with "scale up" and 2) cell death associated with implantation. Large numbers of cells are needed to generate relatively small volumes of tissues. These problems are always associated with the human application of tissue engineering concept. Fortunately, dentistry is the advantageous field because the volume of the newly formed tissue to be needed is relatively small compared with other fields of medicine. However, to ultimately be effective in humans, tissue engineering must generate relatively large volume of tissue, starting with the very few cells. Cell implantation and its associated vascular disruption result in a relatively hypoxic environment and cell death. The potential for different cell types to be expanded *in vitro* and survive a relatively hostile environment at the time of implantation is now being explored. To be effective, cells should be easily procured, be effectively expanded *in vitro* , survive the initial implantation, be accepted as self, function normally, and not become malignant.

Embryonic stem cells (ES cells) and induced pluripotent stem cells (iPS cells) may have very similar potentials and risk to develop into the different cellular elements necessary for effective tissue regeneration. ES cells have been postulated to retain a greater ability to produce healthier tissue despite its ethical issues. At the point in time, there is little evidence that iPS cells can be consistently driven to form only the cell type needed for the tissue to be engineered without any risk of carcinogenicity. Being derived from autogenic cells, iPS cells have the associated problem of malignancy.

It is my belief that some forms of tissue-specific adult stem cells are the most hopeful cell at this moment for clinical use because it may represent Mother Nature's repair cells. Such cells are potentially present within all of the tissue of the body and may remain dormant until they are activated in response to tissue injury. However the chemical environment such as a low oxygen and a poor nutrition at the site of any injury is very hostile. When adequate numbers of cells transplanted into the damaged tissue, the adult stem cell can't survive in this environment. The important thing is the cytokines produced by the transplanted stem cell. They can recruit stem cells from the distant site of whole body. These cytokines change the environment dramatically from sever condition chemically to the adequate condition for tissue repair. Then Mother Nature's repair cells, which are programed to mature and repair tissue damage of a certain magnitude, are activated. Therefore I can conclude that the main player for the tissue regeneration is not stem cells but the cytokines produced by the stem cell. From these speculations we have tried several experiments using the established model animals such as diabetes, hepatitis, stroke, spinal cord injury and so on. The results of the experimental studies were excellent and showed that stem cell based cytokines have almost the same ability as the stem cell for the regeneration in several organs. I believe that we establish a novel approach for tissue and organ regeneration by using stem cell based cytokines.

Minoru Ueda
Department of Oral & Maxillofacial Surgery,
Nagoya University Graduate School of Medicine,
Japan

Chapter 1

Cell Sheet Engineering for Periodontal Regeneration

Takanori Iwata, Kaoru Washio, Toshiyuki Yoshida,
Isao Ishikawa, Tomohiro Ando,
Masayuki Yamato and Teruo Okano

Additional information is available at the end of the chapter

1. Introduction

Periodontitis is a world-wide infectious disease that destroys the tooth-supporting attachment apparatus, which consists of alveolar bone, cementum, and periodontal ligament. Recent studies have reported numerous associations between periodontitis and systemic diseases, such as cardiovascular disease (de Oliveira et al., 2010) and diabetes mellitus (Lalla and Papapanou, 2011), as well as a higher risk of preterm low birth-weight babies (Offenbacher et al., 1996). Furthermore, researches have recently shown that Bisphosphonate-Related Osteonecrosis of the Jaws (BRONJ) is also associated with severe periodontitis (Vescovi et al., 2011). Therefore, periodontal treatment may not only contribute to oral hygiene but also improvement of systemic conditions (Seymour et al., 2007). Conventional treatments, such as scaling, root-planing, and surgical cleaning, have been performed to remove the bacteria and contaminated tissue. However, these procedures frequently result in the formation of a weak attachment, a condition termed "long junctional epithelium (LJE)" (Caton et al., 1980), wherein the patients tend to present with a recurrence of disease without maintenance therapies (Axelsson and Lindhe, 1981). To overcome this problem, various regenerative therapies, such as guided tissue regeneration (GTR) and enamel matrix derivative, have been introduced in clinical practice. The use of cell-occlusive membranes for GTR is regarded as the first generation of periodontal regeneration, whereas the development and use of growth factors and endogenous regenerative technology for periodontal regeneration is regarded as the second generation of periodontal regeneration (Ishikawa et al., 2009). However, the outcomes of these studies were limited and associated with poor clinical predictability (Esposito et al., 2009). Therefore, stem cell-based approaches for periodontal regeneration have been studied and translated into clinical settings as the third generation. In this chapter, we would like to describe the principles of "Cell Sheet Engineering" and its application of clinical settings, featuring our recent translational research for periodontal regeneration.

2. "Cell Sheet Engineering (CSE)"

The cell delivery for periodontal regeneration is usually performed with the combination use of cells and scaffolds, although the location and the differentiation of transplanted is difficult to control. In contrast to approaches that utilize scaffolds, we have developed an alternative technology for cell transplantation using temperature responsive culture dishes, which we call "Cell Sheet Engineering".

2.1. Intelligent surface of *N*-isopropylacrylamide (PIPAAm) and fabrication of cell sheets

Poly(*N*-isopropylacrylamide) (PIPAAm) is a temperature responsive polymer that has been widely utilized for novel biomedical applications. We have developed a PIPAAm-grafted surface as a smart biointerface wherein cell attachment/detachment can be easily controlled by simply changing the temperature (Okano et al., 1995; Yamada et al., 1990). This surface is slightly hydrophobic under cell culture conditions of 37 °C, but readily becomes hydrated and hydrophilic below its lower critical solution temperature (LCST) of 32 °C. Cells can adhere, spread, and proliferate similarly to that on ungrafted tissue culture grade polystyrene surfaces at 37 °C (Figure 1A), and cells detach from the surface by reducing temperature below LCST, making it possible to harvest the cells from the culture surfaces without the use of proteolytic enzymes (Figure 1B).

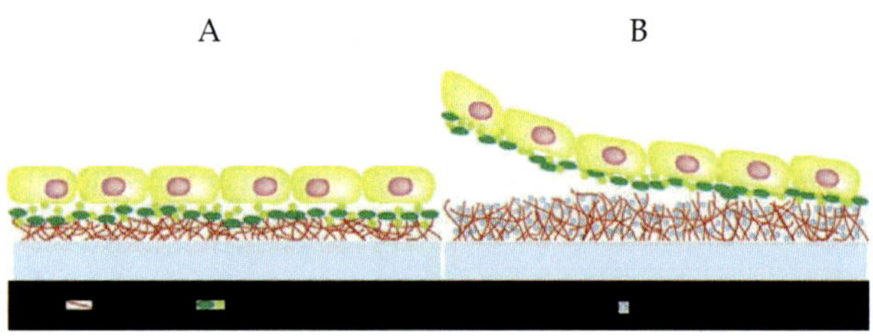

Figure 1. The principle of "Cell Sheet Engineering".

A: Cells can attach and proliferate on grafted surface of the temperature responsive polymer (poly (N-isopropylacrylamide: PIPAAm) at 37 °C, wherein PIPAAm is extensively dehydrated and compact. B: At temperatures below 32 °C, cells with extracellular matrix proteins spontaneously detach from the temperature responsive culture dishes, wherein PIPAAm is fully hydrated with an extended-chain conformation. A simple temperature change can control cell attachment/detachment without any damages. Modified and reprint from Iwata et al., 2013.

The application of this technology has enabled the retrieval of confluently cultured cells, such as keratinocytes (Yamato et al., 2001), corneal epithelial cell sheets (Nishida et al., 2004a), and oral mucosal epithelial cells (Ohki et al., 2006) in the form of a "cell sheet". The epithelial cell sheets are multi-layered and preserve the integrity of proteins such as E-cadherin and laminin 5 that are typically destroyed in the process of enzymatic treatments (Yamato et al., 2001). In addition, recent studies revealed that epithelial cell sheets can be fabricated using temperature responsive culture inserts without feeder layers (Murakami et al., 2006a; b), thereby eliminating exclude xenogeneic factors for animal-free cell transplantation (Takagi et al., 2011).

To fabricate thick tissues, cell sheets can be stacked in layers because they can connect to one another very quickly. A study demonstrated that bilayer cardiomyocyte sheets were completely coupled 46 ± 3 min (mean ± SEM) after the initial layering (Haraguchi et al., 2006), suggesting that multi-layered cell sheets can communicate and become synchronized as functional tissues. Based on this study, multi-layered transplantation was performed (Shimizu et al., 2006b). When more than three cardiomyocyte sheets were layered and transplanted into the subcutaneous space in rats, the appearance of fibrosis and disordered vasculature indicated the presence of fibrotic areas within the transplanted laminar structures. Although the rapid establishment of microvascular networks occurred within the engineered tissues, this formation of new vessels did not rescue the tissues when the thickness was above 80 μm. Using a multiple-step transplantation protocol at 1 or 2 day intervals resulted in rapid neovascularization of the engineered myocardial tissues with a thickness of more than 1 mm (Shimizu et al., 2006b), and these results led us to fabricate prevascularized cell sheets (Sekine et al., 2011). Recent studies demonstrate that the combination of different types of cells, for example an endothelial cell sheet sandwiched with other types of cell sheets, can lead to pre-vascularization *in vitro*, which may allow the graft to survive and function (Haraguchi et al., 2012; Pirraco et al., 2011). Furthermore, the three-dimensional manipulation of fibroblast cell sheets and micro-patterned endothelial cells with a gelatin-coated stacking manipulator produced microvascular-like networks within a 5-day *in vitro* culture (Tsuda et al., 2007). Non-patterned endothelial cell sheets and other types of cell sheets with a fibrin gel manipulator can also produce pre-vascular networks both *in vitro* (Asakawa et al., 2010) and *in vivo* (Sasagawa et al., 2009).

2.2. Cell sheet transplantation in animal models

From the beginning of the 21st century, various types of cells have been extracted, cultured in temperature responsive dishes, and fabricated as cell sheets. Transplantation has been performed, and the efficacy of these cell sheets was evaluated in most of the studies.

2.2.1. Corneal regeneration

Limbal stem-cell deficiency by ocular trauma or diseases causes corneal opacification and loss of vision. To recruit limbal stem cells, a novel cell-sheet manipulation technology that takes advantage of temperature responsive culture surfaces was developed (Nishida et al., 2004a). The results reveal that multi-layered corneal epithelial cell sheets were successfully fabricated and that their characteristics were similar to those of native tissues. Transplantation of these cell sheets induced corneal surface reconstruction in rabbits. For patients who suffer from unilateral limbal stem deficiency, corneal epithelial cell sheets can be cultured from autologous limbal stem cells. When the objective is to repair the bilateral corneal stem cell deficiency, autologous oral mucosal epithelial cells are utilized to create oral mucosal epithelial cell sheets. The cell sheets contain both cell-to-cell junctions and extracellular matrix proteins, and can be transplanted without the use of any carrier substrates or sutures. Therefore, oral mucosal epithelial sheets were examined as an alternative cell source to expand the possibilities of autologous transplantation. Autologous transplantation to rabbit corneal surfaces successfully reconstructed the corneal surface and restored transparency. Four weeks after the transplantation, epithelial stratification was similar to that of normal corneal epithelia, although the keratin expression profile retained characteristics of the oral mucosal epithelium.

2.2.2. Cardiac regeneration

To enhance the function of cardiac tissue, neonatal rat cardiomyocyte sheets were fabricated and examined (Shimizu et al., 2002). When 4 sheets were layered, spontaneous beating of the engineered constructs was observed. When they were transplanted subcutaneously, heart tissue-like structures and neovascularization within the contractile tissues were observed. The long-term survival of pulsatile cardiac grafts was confirmed for more than one year in rats (Shimizu et al., 2006a). Another study was performed to create thick tissue in rats (Shimizu et al., 2006b). However, the thickness limit for the layered cell sheets of subcutaneous tissue was ~80 μm (3 layers). To overcome this limitation, several transplantations of triple-layer grafts were performed, resulting in an approximately 1 mm-thick myocardium with a well-organized microvascular network. Other types of cell sheets were also examined to improve cardiac function. Adipose-derived mesenchymal stem cells in mice (Miyahara et al., 2006) and skeletal myoblasts in dogs, rats, and hamsters (Hata et al., 2006; Hoashi et al., 2009; Kondoh et al., 2006) were transplanted as cell sheets, demonstrating the efficacy of the method for cardiac repair.

2.2.3. Cartilage regeneration

Chondrocyte sheets applicable to cartilage regeneration were prepared using cell sheet manufacturing technique that takes advantage of temperature responsive culture dishes. The layered chondrocyte sheets were able to maintain the phenotype of cartilage and could be attached to sites that exhibited cartilage damage. The cell sheets act as a barrier for preventing the loss of proteoglycan from these sites and for protection against catabolic factors in the joints of rabbits (Kaneshiro et al., 2006).

2.2.4. Esophageal regeneration

With the recent development of endoscopic submucosal dissection (ESD), large esophageal cancers can be removed using a single procedure. However, complications, such as postoperative inflammation and stenosis, frequently occur after an aggressive ESD procedure, which can considerably affect the quality of life of the patient. Therefore, a novel treatment combining ESD and the endoscopic transplantation of tissue-engineered cell sheets created using autologous oral mucosal epithelial cells, was examined in a canine model (Ohki et al., 2006). The results confirm the efficacy of the novel combination of the endoscopic approach with the potential treatment of esophageal cancers that can effectively enhance wound healing and possibly prevent postoperative esophageal stenosis.

2.2.5. Hepatocyte regeneration

To address the demand for therapeutic benefits for patients suffering from liver disease, the development of new therapeutic applications is crucial. Therefore, hepatic tissue sheets transplanted into the subcutaneous space of mice have been investigated, resulting in the efficient engraftment of the surrounding cells, as well as the formation of a two-dimensional hepatic tissues network, which was stable for more than 200 days (Ohashi et al., 2007). The engineered hepatic cell sheets also showed several characteristics of liver-specific functionality, and the use of bilayered sheets enhanced these characteristics.

2.2.6. Fibroblast sheet transplantation for sealing air leaks

In thoracic surgery, the development of postoperative air leaks is the most common cause of prolonged hospitalization. To seal the lung leakage, use of autologous fibroblast sheets on the defects was demonstrated to be an effective treatment for permanently sealing air leaks in a dynamic fashion in rats (Kanzaki et al., 2007). Using roughly the same procedures, pleural defects were also closed by fibroblast sheets in pigs (Kanzaki et al., 2008).

2.2.7. Mesothelial cells for the prevention of post-operative adhesions

Post-operative adhesions often cause severe complications such as bowel obstruction and abdominopelvic pain. The use of mesothelial cell sheets was investigated to prevent post-operative adhesions in a canine model (Asano et al., 2006). Mesothelial cells were harvested from tunica vaginalis (Asano et al., 2005) and cell sheets were fabricated on a fibrin gel. The results demonstrated that mesothelial cell sheets are effective for preventing post-operative adhesion formation.

2.2.8. Retinal Pigment Epithelial (RPE) cell regeneration

The retinal pigment epithelium (RPE) plays an important role in maintaining the health of the neural retina. RPE cell sheets were fabricated as a monolayer structure with intact cell-to-cell junctions, similar to that of native RPE (Kubota et al., 2006). In the transplantation study, RPE cell sheets attached to the host tissues in the subretinal space were more effective than the use of injected isolated cell suspensions in rabbits (Yaji et al., 2009).

2.2.9. Urothelial regeneration

Augmentation cystoplasty using gastrointestinal flaps may induce severe complications such as lithiasis, urinary tract infection, and electrolyte imbalance. The use of viable, contiguous urothelial cell sheets cultured *in vitro* should eliminate these complications. Canine urothelial cell sheets were grown and their structures were shown to be appropriate (Shiroyanagi et al., 2003). Urothelial cell sheets were autografted onto dog demucosalized gastric flaps successfully, with no suturing or fixation, and generated a multi-layered urothelium *in vivo* (Shiroyanagi et al., 2004). The novel intact cell-sheet grafting method rapidly produced native-like epithelium *in vivo*.

2.2.10. Islet regeneration

To establish a novel approach for diabetes mellitus, pancreatic islet cell sheets were fabricated and transplanted in rats (Shimizu et al., 2009). Laminin-5 was coated on temperature responsive dishes to enhance the initial cell attachment, and the presence of specific molecules, such as insulin and glucagon, was also observed in the recipient site.

2.2.11. Thyroid regeneration

For hormonal deficiencies caused by endocrine organ diseases, continuous oral hormone administration is indispensable to supplement the shortage of hormones. To verify the cytotherapeutic approach, cells from rat thyroid were spread on temperature responsive culture dishes, and cell sheets were created (Arauchi et al., 2009). Rats were exposed to total thyroidectomy as hypothyroidism models and received the thyroid cell sheet transplantation 1 week after the total thyroidectomy. The transplantation of the thyroid cell sheets was able to restore the thyroid function 1 week after the cell sheet transplantation and the improvement was observed long after the surgery.

2.3. Cell sheet transplantation in human clinical trials

In Japan, 6 clinical trials using cell sheet engineering technology have been started or have already been completed.

2.3.1. Corneal reconstruction

The first clinical trial of the cell sheet engineering technology involved a corneal reconstruction using autologous mucosal epithelial cells, and the results were published in 2004 (Nishida et al., 2004b). Oral mucosal tissue was harvested from 4 patients with bilateral total corneal stem-cell deficiencies. Subsequently, cells were cultured for two weeks using a mitomycin C-treated 3T3 feeder layer and transplanted directly into the denuded corneal surfaces without sutures. The results demonstrated that complete re-epithelialization of the corneal surfaces occurred, and the vision of all patients was restored. Recently, autologous oral mucosal epithelial cell sheets cultured with UpCell-Insert technology (CellSeed, Tokyo, Japan) without the feeder layer were transplanted into 25 patients for the treatment of corneal limbal epithelial deficiency in France The safety of the products was established during the 360-day follow-up, and the results confirmed its efficacy for reconstructing the ocular surface. (Burillon et al., 2012).

2.3.2. Endoscopic treatment of esophageal ulceration

Using a canine model (Ohki et al., 2006), autologous oral mucosal epithelial cell sheets were fabricated using the UpCell-Insert technology. After performing the esophageal endoscopic submucosal dissection to remove superficial esophageal neoplasms, cell sheets were transplanted, resulting in the complete prevention of stricture formation in patients with partial circumferential resection (Ohki et al., 2009; Ohki et al., 2012).

2.3.3. Improvements in ischemic cardiomyopathy

Autologous myoblast cells from a patient's thigh were fabricated as cell sheets, and these cell sheets were transplanted into end-stage dilated cardiomyopathy patients in need of left ventricular assist systems (Sawa et al., 2012). The myoblastic cell sheets were transplanted into the affected part of the heart in the patients. The first patient was successfully treated and discharged from the hospital without requiring a ventricular assisting device.

2.3.4. Cartilage regeneration

A clinical trial for cartilage regeneration began in 2011 at Tokai University, Japan. In this study, autologous chondrocytes and synoviocytes were co-cultured with the UpCell-Insert technology. After a period of cultivation, co-cultured cell sheets were combined into three layers and transplanted into the cartilage defects of patients.

2.3.5. Nasal mucosa epithelial cell sheet transplantation to the middle ear bone for preventing hearing loss

A clinical trial for preventing hearing loss began in 2014 at The Jikei University, Japan. Autologous nasal mucosa epithelial cell sheets were transplanted to the surface of bone of the middle ear, and inhibit such as the hyperplasy of granulation tissue and bone, and the progression of fibroblast within middle ear cavity, which induce hearing loss after the surgery of otitis media.

3. Periodontal regeneration

Our laboratory started to introduce cell sheet engineering for periodontal regeneration since sometime after 2000. A key event in periodontal regeneration involves the formation of periodontal ligament and cementum complex (MacNeil and Somerman, 1999), which is a thin surface structure that anchors the tooth to the alveolar socket. Several studies have demonstrated that the cell sheet engineering approach can deliver functional cells in the form of a thin layered sheet, wherein the extracellular matrices, cell-cell junctions, and cell-matrix interactions are well-preserved (Kumashiro et al., 2010). Thus, we have attempted to regenerate this periodontal attachment apparatus based on the technology of "cell sheet engineering" (Yang et al., 2007).

3.1. Small animal studies

Human PDL (hPDL) cell sheets were successfully created using temperature responsive dishes, and the characteristics of hPDL cell sheets were investigated (Hasegawa et al., 2005). In this study, explant culture methods were utilized for the primary culture of hPDL cells. The hPDL cell sheets cultured with ascorbic acid were recovered from the culture dishes as a contiguous sheet accompanied by abundant extracellular matrix components, including type I collagen, integrin β1 and fibronectin. Then, hPDL cell sheets were transplanted as cell pellets into a mesial dehiscence model in athymic rats. Four weeks after surgery, newly formed immature fibers with obliquely anchored dentin surfaces were observed in all the experimental sites, whereas no such findings were observed in any control sites (Figure 2).These results suggest that this procedure based upon the principles of cell sheet engineering can be applied to periodontal regeneration.

A B

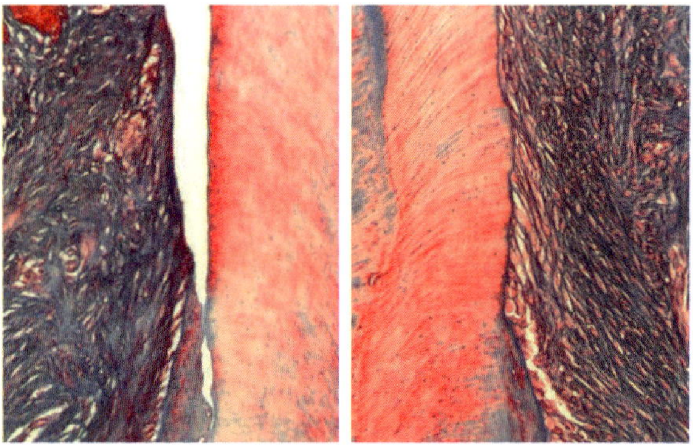

Figure 2. PDL regeneration at 4 weeks postsurgery.

A: Nontransplanted control site. B: hPDL transplanted experimental site. Regeneration of periodontal ligament-like structure was observed only in the experimental site. Azan staining. Modified and reprint from Hasegawa et al., 2005.

Next, the optimal culture condition was examined. Because the osteoinductive medium, which contains 50 μg/ml of ascorbic acid, 10 mM β-glycerophosphate, and 10 nM dexamethasone, enhanced both osteoblastic/cementoblastic and the periodontal differentiation of PDL cells *in vitro*, we compared hPDL cell sheets cultured in the absence and presence of these osteoinductive supplements in a xenogeneic transplantation model (Flores et al., 2008a). Three layered hPDL cell sheets were constructed with fibrin gel and transplanted with a human dentin block into the back of a subcutaneous athymic rat. The constructs were excised for histological investigation 6 weeks after the transplantation. The three-layered hPDL cell sheets-dentin block constructs induced a new cementum-like hard tissue on the surface of the dentin in more than 60% of the samples. Collagen fibers were inserted perpendicularly into the newly formed cementum-like tissue, and this orientation resembled the native Sharpey's fibers. In addition, the regenerative potential of hPDL cell sheets cultured with the osteoinductive medium the was confirmed, when hPDL cell sheets were transplanted onto the root surface of periodontal defects in athymic rat mandibles (Flores et al., 2008b). The results indicate that most of the specimens in the experimental group exhibited a newly-formed cementum and a new attachment of collagen fibers to the cementum layer. No clear cementum layer was observed in the control group (in the absence of osteoinductive supplements). As shown in these experiments, hPDL cells cultured with osteoinductive medium could contribute to the simultaneous regeneration of cementum and PDL.

3.2. Large animal studies

Based on the successful results from small animal studies, we next utilized canine periodontal defect models. Dog PDL (dPDL) cells were extracted using collagenase/dispase digestion. Four individual dPDL cells were successfully isolated and expanded *ex vivo*. Cells were cultured in a standard medium with osteoinductive supplements for 5 days, because longer cultivation induced spontaneous detachment of cell sheets from the UpCell Surfaces. Three-layered dPDL cell sheets were fabiricated with woven polyglycolic acid (PGA) for cell sheet transfer. This PGA product has a number of advantages, including: 1) cell sheets can be easily peeled from temperature responsive dishes, because cell sheets can be attached to the fibers of the woven PGA, 2) the shrinkage of cell sheets can be prevented, 3) easy stacking of multi-layered cell sheets can be achived in a short period of time (see the video attached to the manuscript (Iwata et al., 2009)), 4) easy adjustment of different sizes of cell sheets can be used to cover any defect shape by simply trimming the cell sheets, 5) the ability to make contact on hard tissues and curved surfaces, and 6) the transplant is visible to the operators. dPDL cell sheets were transplanted into the surface of dental roots containing three-wall periodontal defects in an autologous manner, and bone defects were filled with porous beta-tricalcium phosphate (β-TCP). Cell sheet transplantation regenerated both new bone and cementum connecting with the well-oriented collagen fibers, while only limited bone regeneration was observed in the control group where cell sheet transplantation was not performed. These results suggest that PDL cells have multiple differentiation properties that allow for the regeneration of periodontal tissues composed of hard and soft tissues.

Next, we evaluated the safety and efficacy of PDL cell sheets in a one-wall infrabony defect model (Tsumanuma et al., 2011), which is considered to be a severe defect model (Kim et al., 2004). In this study, we also compared the differences in the periodontal healing of various cell sources. PDL cells, bone marrow derived mesenchymal stem cells, and alveolar periosteal cells were obtained from each animal, three-layered canine cell sheets were transplanted in an autologous manner, and bone defects were filled with porous β-TCP with 3% type I collagen gelto stabilize the graft shape. Eight weeks after transplantation, significantly more periodontal regeneration was observed in the newly formed cementum and well-oriented PDL fibers more in the PDL cell sheets group than in the other groups. These results indicate that PDL cell sheets combined with β-TCP/collagen scaffold serve as a promising tool for periodontal regeneration.

3.3. Optimization of human PDL cells

To protect human rights as subjects in clinical trials, the protocol of cytotherapy should be designed based on Good Clinical Practice (GCP) and Good Manufacturing Practice (GMP). Culturing hPDL cells from a single tooth is essential in performing our clinical trial. However, appropriate method for the extraction and expansion of hPDL cells are still not well understood. Thus, we determined the optimal method of isolation and expansion of hPDL cells and then examined their gene expression levels and differentiation potentials, and eventually validated the common characteristics of hPDL cells from 41 samples (Iwata et al., 2010). The hPDL cells were successfully extracted with collagenase/dispase, and then clonal proliferation was performed. Typically, 10 to 100 colonies were observed for a few days after the initial spreading. hPDL cells exhibit the ability to be highly proliferative when cultured at a low cell density. The cells were subcultured for 3 to 4 days, reaching one million cells in 2 weeks. Then, cells were spread on temperature responsive dishes to create a cell sheet in the presence of the osteoinductive medium. Cell sheets were harvested 2 weeks after spreading because the mRNA expression of osteogenic marker genes was strong after that period of time. Quality assurance tests were performed on at least 7 samples, and then the standard phenotypes of hPDL cell sheets were determined.

According to the GCP and GMP guidelines, hPDL cell sheets were created from three healthy volunteer donors at the GMP-grade Cell Processing Center (CPC) in our university (Washio et al., 2010). GMP-grade reagents and certified materials were used for culturing the hPDL cells. The safety and efficacy of "the product (hPDL cell sheets in this case)" was validated for a clinical trials. Prior to performing the cell culture, autologous serum was prepared from the donors. The hPDL cells were cultured under xeno-free conditions, and cell sheets were fabricated using the temperature responsive dishes. Culture sterility was confirmed using conventional tests. Safety was evaluated using the following tests: 1) the soft-agar colony-formation assay, 2) transplantation into nude mice, and 3) the karyotype test (Yoshida et al., 2012). The efficacy of the cell sheets was verified by transplantation with a dentin block into SCID mice. All of these tests revealed that hPDL cell sheets created at the CPC were safe and exhibited the ability to regenerate periodontal tissues. Another set of three hPDL cell sheets from healthy volunteer donors were created at the CPC to optimize the procedures.

3.4. The clinical trial

After approval on the 5th of January 2011, our clinical trial called "Periodontal regeneration with autologous periodontal ligament cell sheets" was initiated to treat patients presenting with the following ailments: 1) infrabony defects with a probing depth of more than 4 mm after the initial therapy, 2) radiographic evidence of infrabony defects, and 3) a redundant tooth that contains healthy periodontal tissue as a cell source. All patients provided written informed consent according to the GCP. Exclusion criteria included the following: 1) relevant medical conditions contraindicating surgical interventions (e.g., diabetes mellitus, cardiovascular, kidney, liver, or lung disease, or compromised immune system), 2) pregnancy or lactation, and 3) heavy tobacco smoking (more than 11 cigarettes a day). The primary outcome of this trial is to evaluate the safety and efficacy of autologous transplantation of periodontal ligament cell sheets. As of the end of May in 2014, 10 cases of autologous PDL cell sheets were transplanted, and the healing process took place uneventfully.

4. Conclusion

The applications of cell sheet engineering for regenerative medicine are mentioned. Various types of cells have been examined and most of them improved the functions of recipients, suggesting that cell sheet engineering can be an alternative strategy for the therapy of tissue engineering. The implementation of robotic systems that allow the safe mass production of sterile cell sheets automatically, as well as further collaboration between researchers and medical professionals will make "cell sheet engineering" the leading edge solution for regenerative medicine (Elloumi-Hannachi et al., 2010).

Acknowledgements

This study was supported by Creation of innovation centers for advanced interdisciplinary research areas Program in the Project for Developing Innovation Systems "Cell Sheet Tissue Engineering Center (CSTEC)" from the Ministry of Education, Culture, Sports, Science and Technology (MEXT), Japan. Teruo Okano is a director of the board of Cell Seed, a stake holder of Cell Seed, and an inventor of cell sheet-related patent. Masayuki Yamato had been a science consultant of Cell Seed until 2011, stake holder and an inventor of cell sheet-related patent. Tokyo Women's Medical University is receiving research fund from CellSeed Inc.

Author details

Takanori Iwata[1,2], Kaoru Washio[1], Toshiyuki Yoshida[1], Isao Ishikawa[1], Tomohiro Ando[2], Masayuki Yamato[1] and Teruo Okano[1]

1 Institute of Advanced Biomedical Engineering and Science, Japan

2 Department of Oral and Maxillofacial Surgery, Tokyo Women's Medical University, Japan

References

[1] Arauchi A, Shimizu T, Yamato M, Obara T, Okano T (2009). Tissue-engineered thyroid cell sheet rescued hypothyroidism in rat models after receiving total thyroidectomy comparing with nontransplantation models. *Tissue Eng Part A* 15(12):3943-3949.

[2] Asakawa N, Shimizu T, Tsuda Y, Sekiya S, Sasagawa T, Yamato M *et al.* (2010). Prevascularization of in vitro three-dimensional tissues created by cell sheet engineering. *Biomaterials* 31(14):3903-3909.

[3] Asano T, Takazawa R, Yamato M, Kageyama Y, Kihara K, Okano T (2005). Novel and simple method for isolating autologous mesothelial cells from the tunica vaginalis. *BJU Int* 96(9):1409-1413.

[4] Asano T, Takazawa R, Yamato M, Takagi R, Iimura Y, Masuda H *et al.* (2006). Transplantation of an autologous mesothelial cell sheet prepared from tunica vaginalis prevents post-operative adhesions in a canine model. *Tissue Eng* 12(9):2629-2637.

[5] Axelsson P, Lindhe J (1981). The significance of maintenance care in the treatment of periodontal disease. *Journal of Clinical Periodontology* 8(4):281-294.

[6] Burillon C, Huot L, Justin V, Nataf S, Chapuis F, Decullier E *et al.* (2012). Cultured autologous oral mucosal epithelial cell sheet (CAOMECS) transplantation for the treatment of corneal limbal epithelial stem cell deficiency. *Investigative ophthalmology & visual science* 53(3):1325-1331.

[7] Caton J, Nyman S, Zander H (1980). Histometric evaluation of periodontal surgery. II. Connective tissue attachment levels after four regenerative procedures. *J Clin Periodontol* 7(3):224-231.

[8] de Oliveira C, Watt R, Hamer M (2010). Toothbrushing, inflammation, and risk of cardiovascular disease: results from Scottish Health Survey. *Bmj* 340(c2451.

[9] Elloumi-Hannachi I, Yamato M, Okano T (2010). Cell sheet engineering: a unique nanotechnology for scaffold-free tissue reconstruction with clinical applications in regenerative medicine. *Journal of Internal Medicine* 267(1):54-70.

[10] Esposito M, Grusovin MG, Papanikolaou N, Coulthard P, Worthington HV (2009). Enamel matrix derivative (Emdogain(R)) for periodontal tissue regeneration in intrabony defects. *Cochrane database of systematic reviews* 4):CD003875.

[11] Flores MG, Hasegawa M, Yamato M, Takagi R, Okano T, Ishikawa I (2008a). Cementum-periodontal ligament complex regeneration using the cell sheet technique. *J Periodontal Res* 43(3):364-371.

[12] Flores MG, Yashiro R, Washio K, Yamato M, Okano T, Ishikawa I (2008b). Periodontal ligament cell sheet promotes periodontal regeneration in athymic rats. *J Clin Periodontol* 35(12):1066-1072.

[13] Haraguchi Y, Shimizu T, Yamato M, Kikuchi A, Okano T (2006). Electrical coupling of cardiomyocyte sheets occurs rapidly via functional gap junction formation. *Biomaterials* 27(27):4765-4774.

[14] Haraguchi Y, Shimizu T, Sasagawa T, Sekine H, Sakaguchi K, Kikuchi T *et al.* (2012). Fabrication of functional three-dimensional tissues by stacking cell sheets in vitro. *Nat Protoc* 7(5):850-858.

[15] Hasegawa M, Yamato M, Kikuchi A, Okano T, Ishikawa I (2005). Human periodontal ligament cell sheets can regenerate periodontal ligament tissue in an athymic rat model. *Tissue Eng* 11(3-4):469-478.

[16] Hata H, Matsumiya G, Miyagawa S, Kondoh H, Kawaguchi N, Matsuura N *et al.* (2006). Grafted skeletal myoblast sheets attenuate myocardial remodeling in pacing-induced canine heart failure model. *J Thorac Cardiovasc Surg* 132(4):918-924.

[17] Hoashi T, Matsumiya G, Miyagawa S, Ichikawa H, Ueno T, Ono M *et al.* (2009). Skeletal myoblast sheet transplantation improves the diastolic function of a pressure-overloaded right heart. *J Thorac Cardiovasc Surg* 138(2):460-467.

[18] Ishikawa I, Iwata T, Washio K, Okano T, Nagasawa T, Iwasaki K *et al.* (2009). Cell sheet engineering and other novel cell-based approaches to periodontal regeneration. *Periodontol 2000* 51(220-238.

[19] Iwata T, Yamato M, Tsuchioka H, Takagi R, Mukobata S, Washio K *et al.* (2009). Periodontal regeneration with multi-layered periodontal ligament-derived cell sheets in a canine model. *Biomaterials* 30(14):2716-2723.

[20] Iwata T, Yamato M, Zhang Z, Mukobata S, Washio K, Ando T *et al.* (2010). Validation of human periodontal ligament-derived cells as a reliable source for cytotherapeutic use. *Journal of Clinical Periodontology* 37(12):1088-1099.

[21] Iwata T, Washio K, Yoshida T, Ishikawa I, Ando T, Yamato M *et al.* (2013). Cell sheet engineering and its application for periodontal regeneration. *J Tissue Eng Regen Med.*

[22] Kaneshiro N, Sato M, Ishihara M, Mitani G, Sakai H, Mochida J (2006). Bioengineered chondrocyte sheets may be potentially useful for the treatment of partial thickness defects of articular cartilage. *Biochem Biophys Res Commun* 349(2):723-731.

[23] Kanzaki M, Yamato M, Yang J, Sekine H, Kohno C, Takagi R *et al.* (2007). Dynamic sealing of lung air leaks by the transplantation of tissue engineered cell sheets. *Biomaterials* 28(29):4294-4302.

[24] Kanzaki M, Yamato M, Yang J, Sekine H, Takagi R, Isaka T *et al.* (2008). Functional closure of visceral pleural defects by autologous tissue engineered cell sheets. *Eur J Cardiothorac Surg* 34(4):864-869.

[25] Kim CS, Choi SH, Chai JK, Cho KS, Moon IS, Wikesjo UM *et al.* (2004). Periodontal repair in surgically created intrabony defects in dogs: influence of the number of bone walls on healing response. *Journal of Periodontology* 75(2):229-235.

[26] Kondoh H, Sawa Y, Miyagawa S, Sakakida-Kitagawa S, Memon IA, Kawaguchi N *et al.* (2006). Longer preservation of cardiac performance by sheet-shaped myoblast implantation in dilated cardiomyopathic hamsters. *Cardiovasc Res* 69(2):466-475.

[27] Kubota A, Nishida K, Yamato M, Yang J, Kikuchi A, Okano T *et al.* (2006). Transplantable retinal pigment epithelial cell sheets for tissue engineering. *Biomaterials* 27(19):3639-3644.

[28] Kumashiro Y, Yamato M, Okano T (2010). Cell attachment-detachment control on temperature-responsive thin surfaces for novel tissue engineering. *Ann Biomed Eng* 38(6):1977-1988.

[29] Lalla E, Papapanou PN (2011). Diabetes mellitus and periodontitis: a tale of two common interrelated diseases. *Nat Rev Endocrinol* 7(12):738-748.

[30] MacNeil RL, Somerman MJ (1999). Development and regeneration of the periodontium: parallels and contrasts. *Periodontology 2000* 19(1):8-20.

[31] Miyahara Y, Nagaya N, Kataoka M, Yanagawa B, Tanaka K, Hao H *et al.* (2006). Monolayered mesenchymal stem cells repair scarred myocardium after myocardial infarction. *Nat Med* 12(4):459-465.

[32] Murakami D, Yamato M, Nishida K, Ohki T, Takagi R, Yang J *et al.* (2006a). The effect of micropores in the surface of temperature-responsive culture inserts on the fabrication of transplantable canine oral mucosal epithelial cell sheets. *Biomaterials* 27(32): 5518-5523.

[33] Murakami D, Yamato M, Nishida K, Ohki T, Takagi R, Yang J *et al.* (2006b). Fabrication of transplantable human oral mucosal epithelial cell sheets using temperature-responsive culture inserts without feeder layer cells. *J Artif Organs* 9(3):185-191.

[34] Nishida K, Yamato M, Hayashida Y, Watanabe K, Maeda N, Watanabe H *et al.* (2004a). Functional bioengineered corneal epithelial sheet grafts from corneal stem cells expanded ex vivo on a temperature-responsive cell culture surface. *Transplantation* 77(3):379-385.

[35] Nishida K, Yamato M, Hayashida Y, Watanabe K, Yamamoto K, Adachi E *et al.* (2004b). Corneal reconstruction with tissue-engineered cell sheets composed of autologous oral mucosal epithelium. *N Engl J Med* 351(12):1187-1196.

[36] Offenbacher S, Katz V, Fertik G, Collins J, Boyd D, Maynor G *et al.* (1996). Periodontal infection as a possible risk factor for preterm low birth weight. *Journal of Periodontology* 67(10 Suppl):1103-1113.

[37] Ohashi K, Yokoyama T, Yamato M, Kuge H, Kanehiro H, Tsutsumi M *et al.* (2007). Engineering functional two- and three-dimensional liver systems in vivo using hepatic tissue sheets. *Nat Med* 13(7):880-885.

[38] Ohki T, Yamato M, Murakami D, Takagi R, Yang J, Namiki H *et al.* (2006). Treatment of oesophageal ulcerations using endoscopic transplantation of tissue-engineered autologous oral mucosal epithelial cell sheets in a canine model. *Gut* 55(12):1704-1710.

[39] Ohki T, Yamato M, Ota M, Murakami D, Takagi R, Kondo M *et al.* (2009). Endoscopic Transplantation of Human Oral Mucosal Epithelial Cell Sheets-World's First Case of Regenerative Medicine Applied to Endoscopic Treatment. *Gastrointestinal Endoscopy* 69(5):AB253-AB254.

[40] Ohki T, Yamato M, Ota M, Takagi R, Murakami D, Kondo M *et al.* (2012). Prevention of esophageal stricture after endoscopic submucosal dissection using tissue-engineered cell sheets. *Gastroenterology* 143(3):582-588 e581-582.

[41] Okano T, Yamada N, Okuhara M, Sakai H, Sakurai Y (1995). Mechanism of cell detachment from temperature-modulated, hydrophilic-hydrophobic polymer surfaces. *Biomaterials* 16(4):297-303.

[42] Pirraco RP, Obokata H, Iwata T, Marques AP, Tsuneda S, Yamato M *et al.* (2011). Development of osteogenic cell sheets for bone tissue engineering applications. *Tissue engineering Part A* 17(11-12):1507-1515.

[43] Sasagawa T, Shimizu T, Sekiya S, Haraguchi Y, Yamato M, Sawa Y *et al.* (2009). Design of prevascularized three-dimensional cell-dense tissues using a cell sheet stacking manipulation technology. *Biomaterials.*

[44] Sawa Y, Miyagawa S, Sakaguchi T, Fujita T, Matsuyama A, Saito A *et al.* (2012). Tissue engineered myoblast sheets improved cardiac function sufficiently to discontinue LVAS in a patient with DCM: report of a case. *Surg Today* 42(2):181-184.

[45] Sekine H, Shimizu T, Dobashi I, Matsuura K, Hagiwara N, Takahashi M *et al.* (2011). Cardiac cell sheet transplantation improves damaged heart function via superior cell survival in comparison with dissociated cell injection. *Tissue engineering Part A* 17(23-24):2973-2980.

[46] Seymour GJ, Ford PJ, Cullinan MP, Leishman S, Yamazaki K (2007). Relationship between periodontal infections and systemic disease. *Clin Microbiol Infect* 13 Suppl 4(3-10.

[47] Shimizu H, Ohashi K, Utoh R, Ise K, Gotoh M, Yamato M *et al.* (2009). Bioengineering of a functional sheet of islet cells for the treatment of diabetes mellitus. *Biomaterials* 30(30):5943-5949.

[48] Shimizu T, Yamato M, Isoi Y, Akutsu T, Setomaru T, Abe K *et al.* (2002). Fabrication of pulsatile cardiac tissue grafts using a novel 3-dimensional cell sheet manipulation technique and temperature-responsive cell culture surfaces. *Circ Res* 90(3):e40.

[49] Shimizu T, Sekine H, Isoi Y, Yamato M, Kikuchi A, Okano T (2006a). Long-term survival and growth of pulsatile myocardial tissue grafts engineered by the layering of cardiomyocyte sheets. *Tissue Eng* 12(3):499-507.

[50] Shimizu T, Sekine H, Yang J, Isoi Y, Yamato M, Kikuchi A *et al.* (2006b). Polysurgery of cell sheet grafts overcomes diffusion limits to produce thick, vascularized myocardial tissues. *FASEB J* 20(6):708-710.

[51] Shiroyanagi Y, Yamato M, Yamazaki Y, Toma H, Okano T (2003). Transplantable urothelial cell sheets harvested noninvasively from temperature-responsive culture surfaces by reducing temperature. *Tissue Eng* 9(5):1005-1012.

[52] Shiroyanagi Y, Yamato M, Yamazaki Y, Toma H, Okano T (2004). Urothelium regeneration using viable cultured urothelial cell sheets grafted on demucosalized gastric flaps. *BJU Int* 93(7):1069-1075.

[53] Takagi R, Yamato M, Murakami D, Kondo M, Yang J, Ohki T *et al.* (2011). Preparation of keratinocyte culture medium for the clinical applications of regenerative medicine. *Journal of tissue engineering and regenerative medicine* 5(4):e63-73.

[54] Tsuda Y, Shimizu T, Yamato M, Kikuchi A, Sasagawa T, Sekiya S *et al.* (2007). Cellular control of tissue architectures using a three-dimensional tissue fabrication technique. *Biomaterials* 28(33):4939-4946.

[55] Tsumanuma Y, Iwata T, Washio K, Yoshida T, Yamada A, Takagi R *et al.* (2011). Comparison of different tissue-derived stem cell sheets for periodontal regeneration in a canine 1-wall defect model. *Biomaterials* 32(25):5819-5825.

[56] Vescovi P, Campisi G, Fusco V, Mergoni G, Manfredi M, Merigo E *et al.* (2011). Surgery-triggered and non surgery-triggered Bisphosphonate-related Osteonecrosis of the Jaws (BRONJ): A retrospective analysis of 567 cases in an Italian multicenter study. *Oral Oncol* 47(3):191-194.

[57] Washio K, Iwata T, Mizutani M, Ando T, Yamato M, Okano T *et al.* (2010). Assessment of cell sheets derived from human periodontal ligament cells: a pre-clinical study. *Cell and Tissue Research* 341(3):397-404.

[58] Yaji N, Yamato M, Yang J, Okano T, Hori S (2009). Transplantation of tissue-engineered retinal pigment epithelial cell sheets in a rabbit model. *Biomaterials* 30(5): 797-803.

[59] Yamada N, Okano T, Sakai H, Karikusa F, Sawasaki Y, Sakurai Y (1990). Thermo-responsive polymeric surfaces; control of attachment and detachment of cultured cells. *Die Makromolekulare Chemie, Rapid Communications* 11(11):571-576.

[60] Yamato M, Utsumi M, Kushida A, Konno C, Kikuchi A, Okano T (2001). Thermo-responsive culture dishes allow the intact harvest of multilayered keratinocyte sheets without dispase by reducing temperature. *Tissue Eng* 7(4):473-480.

[61] Yang J, Yamato M, Shimizu T, Sekine H, Ohashi K, Kanzaki M *et a*l. (2007). Reconstruction of functional tissues with cell sheet engineering. Biomaterials 28(34): 5033-5043.

[62] Yoshida T, Washio K, Iwata T, Okano T, Ishikawa I (2012). Current status and future development of cell transplantation therapy for periodontal tissue regeneration. Int J Dent 2012(307024.

Chapter 2

The Cell-Multilayered Periosteal Sheet — A Promising Osteogenic and Osteoinductive Grafting Material

Tomoyuki Kawase, Kazuhiro Okuda,
Masaki Nagata and Hiromasa Yoshie

Additional information is available at the end of the chapter

http://dx.doi.org/10.5772/58902

1. Introduction

The present is an exciting time in the field of regenerative biology in dental medicine. Many basic and clinical studies have been performed to improve periodontal therapy and rehabilitation therapy for morphological and functional recovery of the maxillofacial bones. Many materials, e.g., hydroxyapatite, GTR membrane, enamel matrix derivative, and platelet-rich plasma, have been applied to osseous defects to achieve significant outcomes. In contrast, cell therapy, as an advanced tissue-engineering technology, using bone marrow stem cells, periosteal cells, or periodontal ligament cells have just been started in several university hospitals and institutes in Japan.

The periosteum has been recognized as a promising source of immature osteogenic progenitor cells. Osteogenic cells are important components as a promoter of bone regenerative therapy because their bioactivities are crucial to reconstituting an active site of controlled bone formation. We demonstrated that implantation of cultured periosteal sheets into nude mice could induce the formation of bone-like tissue [Kawase *et al.*, 2009]. In addition, we established regenerative surgery with autologous cultured periosteal sheets in combination with platelet-rich plasma (PRP) and hydroxyapatite (HA) in periodontitis patients (Fig. 1) [Yamamiya *et al.*, 2008, Okuda *et al.*,2013]. Furthermore, we reported remarkable bone remodeling in reduced alveolar bone from autologous bone grafting with cultured periosteal sheets [Nagata *et al.*, 2012].

In this chapter, we will introduce the characteristics of periosteal sheets and clinical outcome by cell-based therapy in patients with periodontitis and alveolar jaw reduction. In addition, future perspectives on regenerative therapy will be described.

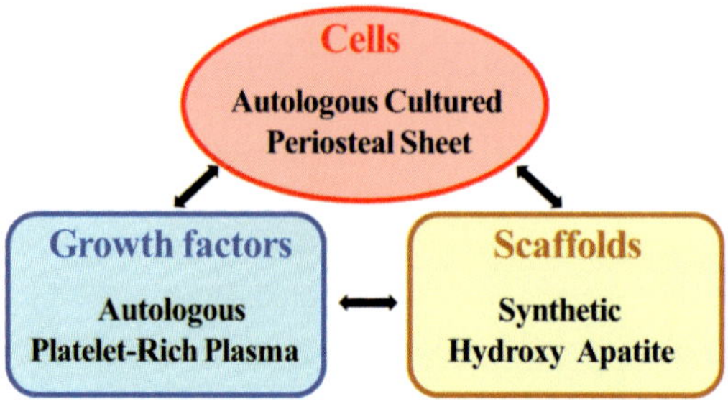

Figure 1. Cultured periosteal sheet application to infrabony defects in patients with periodontitis

2. Biological characteristics of the cultured periosteal sheet

Our periosteal sheets are prepared using static explant cultures of small pieces of alveolar periosteum tissue. The number of cells can be expanded at least by 1000 times by the end of the 6-week preparation, and the resulting periosteal sheet is mechanically tough enough to be picked up by forceps [Kawase *et al.*, 2009]. In our experience preparing more than 2000 periosteal sheets, we confirmed that the growth is not significantly dependent either on age or sex. The successful preparation is rather dependent on the cell density in excised periosteum tissue segments. It is also important to handle the excised periosteum tissue in a timely manner, especially in the initial preparation of the "cut and paste" onto culture dishes because this procedure includes a time-sensitive semi-drying process for facilitating tissue segment attachment onto the plastic surface.

We developed this grafting material as a substitute for crushed autologous bone, a gold standard osteogenic grafting material. Therefore, the periosteal sheets are expected to express osteogenic and osteoinductive capabilities besides (or rather than) the osteoconductive capability displayed by most recent bone filler made of bioactive ceramics.

In our basic studies using in vitro cell culture systems and in vivo animal implantation systems, we have demonstrated several remarkable characteristics of the periosteal sheet, which are summarized below:

Tissue-like thickness: The periosteal sheet displays a unique structure that is composed of cell-multilayers and abundantly deposited extracellular matrices (ECM).

Osteogenicity: Periosteal cells within the periosteal sheet differentiate to osteogenic cells slowly but spontaneously with time of expansion using a conventional medium.

Osteoinduction: The periosteal sheet produces the major growth factors involved in bone metabolism.

2.1. Tissue-like thickness

Because of the cell-multilayered structure, as shown in Figure 2 (Left), the periosteal sheet is substantially thicker than a monolayer cell sheet [Kawase *et al.*, 2009]. In addition, the thickness of the periosteal sheet can be enhanced by many types of commercially available stem-cell culture media [Uematsu *et al.*, 2013a, 2013b] (Fig. 2 Right). However, intact periosteum tissue is hardly excised for expansion culture, and the native periosteum-like structure, dual-layers composed of the cambium and fibrous layers, could not be newly formed or regenerated in vitro under our static culture expansion conditions.

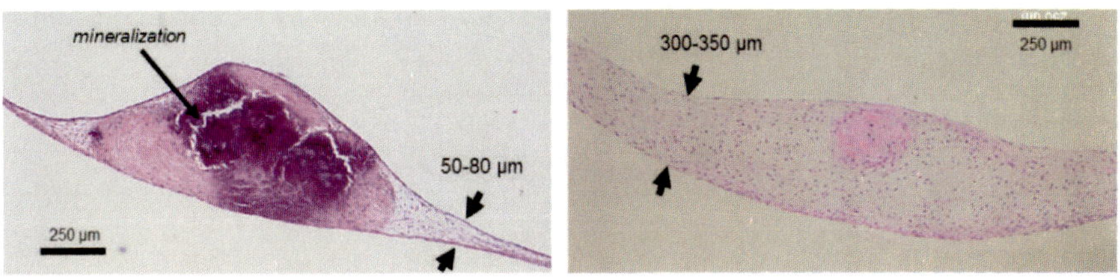

Medium 199 + 10% FBS at Day 35 | Stem cell medium (MesenCult) at Day 35

Figure 2. Effects of culture media on the morphology of periosteal sheets.

Periosteal sheets were expanded with the conventional growth medium (Medium 199 supplemented with 10% FBS) or a commercial stem-cell medium (MesenCult) for 35 days. Saggital paraffin sections were stained with hematoxylin & eosin.

Interestingly, as far as we could attempt and achieve, similar cell-multilayered structures can barely be formed by similar expansion cultures of other connective tissues, such as epidermal and periodontal ligament tissue [Kawase *et al.*, Unpublished observations]. It is also impossible to form similar structures using dispersed single periosteal cell populations in conventional two-dimensional (2D) culture systems. However, using appropriate scaffolding materials, we developed a 3D culture system for preparation of similar periosteal sheets [Kawase *et al.*, 2014].

Material	Region	Stiffness	n
Periosteal sheet	central	5.59 ± 3.24 kPa	76
Periosteal sheet	peripheral	4.35 ± 2.69 kPa	63
Dispersed periosteal cells		14.3 ± 9.42 kPa	118
Dispersed periosteal cells (living)		9.04 ± 9.20 kPa	54
Tissue culture polystyrene		2.34 ± 1.54 GPa	79

Stiffness was evaluated by the nanoindentation technique using an atomic force microscopy.

Table 1. The apparent stiffness of periosteal sheets.

On the other hand, from a biomechanical point of view, ECM-enriched structures function as a soft scaffolding material for proliferating periosteal cells. As shown in Table 1, the stiffness of a conventional plastic dish is several GPa, whereas that of a light-fixed periosteal sheet is 4-6 kPa [Horimizu *et al.*, 2013]. Because stiffer materials are in general more suitable for cell mitosis and proliferation [Vincent *et al.*, 2013], it can be predicted that conventional culture dishes provide more preferable places for periosteal cells to proliferate. This is most likely a major reason why dispersed periosteal cells grow faster than cells contained in a periosteal sheet. It has also been reported that the optical stiffness for the osteoblastic differentiation is 25-40 kPa, and softer material surfaces maintain cells in the osteoblast lineage at immature stages [Engler *et al.*, 2006]. Therefore, it is presumable that a thicker periosteal sheet could be suitable for the expansion of immature osteoprogenitor cells without inducing the osteoblastic differentiation.

2.2. Osteogenicity

As a promising osteogenic grafting material, a periosteal sheet should produce mineral deposits in vitro and form bone-like tissue in vivo. In conventional growth medium, Medium 199 supplemented with 10% fetal bovine serum (FBS), periosteal cells gradually and increasingly express alkaline phosphatase (ALP) with expansion time [Kawase *et al.*, 2009]. The addition of chemical agents involved in osteoblastic differentiation, such as dexamethasone, β-glycerophosphate, and ascorbate, immediately increases ALP activity and simultaneously mineral deposit formation [Kawase *et al.*, 2009, 2010, 2011].

When implanted into the subcutaneous tissue of nude mice, these periosteal sheets, regardless of their differentiation stages, are capable of forming osteoid and bone-like tissue more potently than dispersed periosteal cells combined with porous hydroxyapatite blocks [Kawase *et al.*, 2009, Uematsu *et al.*, 2013a]. In addition, several periosteal cells, especially those embedded in dense collagen matrices, become osteocyte-like cells. These data suggest the "osteogenicity" of the periosteal sheet.

2.3. Osteoinduction

We have detected major growth factors, such as several interleukins, FGF-2 (fibroblast growth factor-2), MCP-1 (monocyte chemotactic protein-1), TPO (thrombopoeitin), IGFBP-2 (insulin-like growth factor binding protein-2), GRO (growth-related oncogene) and VEGF (vascular endothelial growth factor) in conditioned media of the periosteal sheet [Kawase *et al.*, 2009, Kobayashi *et al.*, Manuscript in submission]. The osteogenic induction using chemical agents drastically changed the profile of cytokine release, as shown in Table 2.

Following implantation of periosteal sheets into the subcutaneous tissue of nude mice, global DNA microarray analyses demonstrated that the periosteal sheet influences the surrounding host tissue and induces osteoclastic phenotypes (Table 3). We also observed that the osteoblastic phenotypes could be induced by the implanted periosteal sheet. In addition, the osteoinduced periosteal sheet is potently capable of inducing angiogenesis and osteoclastic formation. Therefore, these data suggest "osteoinduction" by the periosteal sheet.

Changes	Growth factors
control > differentiated	
moderate	IL-6 MCP-1, TPO
control < differentiated	
moderate	GRO-α, IL-8, IL-13, IGFBP-2
substantial	GRO, IL-5, IL-7, IL-10, MCP-3

Growth factors released by the control and the osteoinduced periosteal sheets to the low-serum culture medium were semi-quantitatively evaluated using an antibody array.

Table 2. Effects of the osteoinduction on the growth factor production by the periosteal sheet.

Phenotype	Genes
Upregulation (> twofolds)	
Osteoblast	RUNX2, SatB2, Osteoprotegerin, BMP1, TGFβ1, integrin α5β1, ICAM1
Osteoclast	TRAP, Cathepsin K, Osteoclast-like cell cDNA
Blood vessel	Tie1
Downregulation (> twofolds)	
Stem cell	SOX2
Chondrocyte	SOX9, COL2
Osteoblast	MSX2, ALP, BMP4, Bone gla protein, IGF2
Osteoclast	CSF1

The human periosteal sheets, which were implanted into the subcutaneous tissue of nude mice, were retrieved along with the surrounding mouse tissue. Mouse-specific mRNAs expressed in the surrounding tissue were evaluated by an Agilent Mouse Genome (4 x 44K) microarray.

Table 3. Global DNA microarray analyses of gene expression in the surrounding mouse connective tissue.

2.4. Conclusion

We should again emphasize that the periosteal sheet is characterized by its unique structure. Dispersed periosteal cells and other cells in the osteoblast lineage expanded in 2D culture systems are capable of producing growth factors involved in bone metabolism and forming mineral deposits as well as the periosteal sheet. However, these cells require appropriate scaffolding materials for implantation. In addition, a 3D culture system without using appropriate scaffolding materials, such as accumulation of cell sheets and large-scale spheroid formations, typically induces cell necrosis in the central regions. Abundant ECM functions as a scaffolding material and avoids the increase in cell density, preventing inevitable necrosis.

3. Application of cultured periosteal sheet for periodontal regeneration

To regenerate periodontal tissue destroyed by chronic periodontitis, human autologous cultured periosteal sheets in combination with platelet-rich plasma (PRP) [Okuda *et al.*, 2003] and hydroxyapatite (HA) granules, were clinically applied. The periosteum specimens were dissected from each patient and were incubated for approximately six weeks until the sheets were formed. Prior to placement of the cultured periosteal sheets, osseous defects were thoroughly debrided and filled with HA granules in combination with PRP. Standardized clinical and radiographic measurements were performed during follow-up examinations at one and five years. Clinical attachment level (CAL) and radiographic infrabony defect depth (IBD) were determined as the primary and secondary end-points.

The one-year results of this treatment indicate favorable clinical results, when compared to baseline or when compared to the control group (PRP with HA). A five-year follow-up evaluation of 22 selected patients after cultured periosteal sheets treatment indicated the treated infrabony defects remained clinically stable and were well maintained. Radiographically, there was increased osseous radiopacity and bone trabeculation, suggesting further bone maturation.

A factor likely contributing to these favorable clinical results is the presence of osteogenic cells in the cultured periosteal sheets, which provided greater regeneration potential.

3.1. Case presentation

In this case report, clinical and radiographic evaluation of the clinically representative case #22 of a cultured periosteal sheet in combination with a PRP and HA granule mixture surgically treated site is presented in Figure 3.

A 56-year-old Japanese female presented with clinical and radiographic (Fig. 3A) evidence of chronic periodontitis and bone loss on the mesial surface of the mandibular left first molar. At baseline, the PD was 6 mm, and the CAL was 6 mm. Upon surgical exposure, a 5-mm occlusal-apical by 5-mm mesio-distal three-walled infrabony osseous defect was revealed on the mesial surface of the molar (Fig. 3B). After debridement and thorough root planing, a PRP and HA granule complex was placed into the osseous infrabony defect (Fig. 3C) and overlaid with a cultured periosteal sheet (Fig. 3D).

At 1-year post-surgery, the PD and CAL were 3 mm and 5 mm, respectively, representing a 1-mm gain in CAL from baseline, and this favorable clinical response in the decrease of PD and gain in CAL was maintained at 5 years (Fig. 3E). Increased radiopacity was demonstrated on the mesial surface of the molar at 1 year and 5 years (Fig. 3F) [Okuda *et al.*, 2009, 2013].

http://dx.doi.org/10.5772/58902

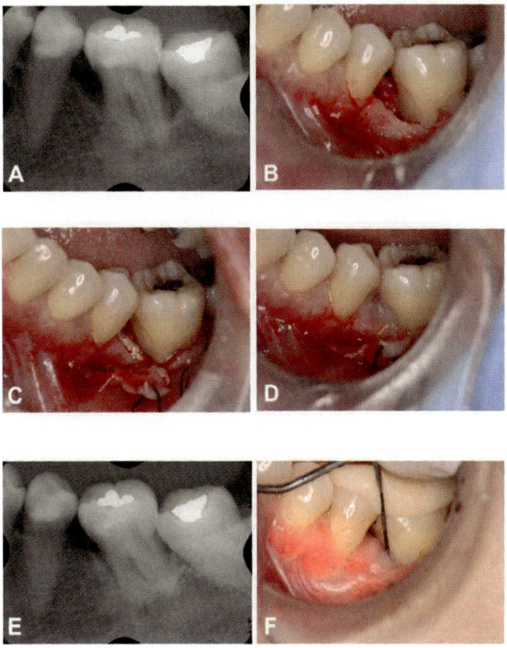

Figure 3. Clinical and radiographic appearance during follow-up examinations at baseline and five years. A: Surgical area before surgery with IBD depicting 5 mm of bone loss, B: Reflection and debridement of the area revealed a three-walled osseous defect, C: Grafting with PRP+HA granules, D: The surgical site following placement of the cultured periosteal sheet covering the PRP+HA granules, E: Clinical view at 5-year follow-up, F: Radiographs at 5 years showing that the cortical bony architecture and periodontal ligament space appear normal.

3.2. Case-controlled clinical trial

Thirty interproximal infrabony osseous defects in 30 healthy, non-smoking subjects diagnosed with chronic periodontitis were included in this study with one infrabony defect with PD ≥ 6 mm, CAL ≥ 6 mm, and an osseous defect depth estimated to be ≥3 mm when measured radiographically. The subjects were randomly assigned to a test group (cultured periosteal sheets combined with PRP and HA) or a control group (PRP with HA). Clinical and radiographic measurements were performed at baseline and the 1-year post-surgical evaluation. Compared to baseline, the 1-year results indicated that both treatment modalities resulted in significant changes ($P < 0.01$) in the probing depth, clinical attachment level, and radiographic infrabony defect depth. Compared to the control group, the test group exhibited a significantly more favorable change in clinical attachment (4.1 ± 1.2 mm versus 5.3 ± 1.5 mm; $P < 0.05$) and radiographic infrabony defect (0.4 ± 0.9 mm versus 1.7 ± 1.2 mm; $P < 0.01$), as shown in Table 4 [Okuda *et al.*, 2005, Yamamiya *et al.*, 2008].

Clinical and Radiographic Index and Treatment Group	Baseline	1-year	*P* value*
PD (mm)			
Periosteal sheet + PRP + HA (Test)	7.7 ± 1.1	2.9 ± 0.4	$P < 0.01$
PRP + HA (Control)	7.6 ± 1.1	3.3 ± 1.0	$P < 0.01$
P value†	NS	NS	
CAL (mm)			
Periosteal sheet + PRP + HA	8.1 ± 1.2	4.1 ± 1.2	$P < 0.01$
PRP + HA	8.0 ± 1.3	5.3 ± 1.5	$P < 0.01$
P value	NS	$P < 0.05$	
IBD (mm)			
Periosteal sheet + PRP + HA	5.3 ± 1.1	0.4 ± 0.9	$P < 0.01$
PRP + HA	4.9 ± 1.4	1.7 ± 1.2	$P < 0.01$
P value	NS	$P < 0.01$	

Means ± SD (n=15 subjects in each treatment group). SD=standard deviation. NS=not significant, $P > 0.05$, * The significance level between baseline and 1-year, † Statistical significance level between the study groups at baseline or 1-year.

Table 4. Mean clinical and radiographic measurements at baseline and 1 year for the test (periosteal sheet+PRP+HA) and control (PRP+HA) study groups.

3.3. Five-year follow-up study

Twenty-two patients who had received surgery using cultured periosteal sheets were included in the study for statistical analysis comparing the 1-year results and the 5-year follow-up results. Significant differences between baseline and the 1-and 5-year data were evaluated by means of the Wilcoxon signed-rank matched-pair test, taking into account the paired nature of the observations. The mean PD, CAL and IBD scores at baseline are as follows: PD, 7.6 ± 1.4 mm; CAL, 8.0 ± 1.4 mm and IBD, 4.6 ± 1.4 mm (Table 5).

The 1-year results after treatment indicated the mean PD, CAL and IBD values were improved significantly as compared with baseline (PD: 2.9 ± 0.4 mm at 1-year versus 7.6 ± 1.4 mm at baseline, $P < 0.01$; CAL: 4.8 ± 1.4 mm at 1-year versus 8.0 ± 1.4 mm at baseline, $P < 0.01$; IBD: 1.4 ± 1.1 mm at 1-year versus 4.6 ± 1.0 mm at baseline, $P < 0.01$).

At the 5-year follow-up, as compared to the 1-year evaluation period, the mean PD and CAL were stable (PD: 2.9 ± 0.3 mm at 5-years *vs.* 2.9 ± 0.4 mm at 1 year, NS, CAL: 4.6 ± 1.2 mm at 5-year *vs.* 4.8 ± 1.4 mm at 1-year; NS), and IBD was improved significantly when compared to the 1-year result (0.2 ± 0.5 mm at 5-years *vs.* 1.4 ± 1.1 mm at 1-year, $P < 0.01$). It is also noteworthy that there were no infectious episodes and no other adverse complications associated with treatment over the five-year time period of this study [Okuda *et al.*, 2013].

	Baseline	1-year	5-years	*p value*		
				1 vs BL*	5 vs BL†	5 vs 1‡
Clinical Parameter						
PD	7.6 ± 1.4	2.9 ± 0.4	2.9 ± 0.3	$P < 0.01$	$P < 0.01$	NS
CAL	8.0 ± 1.4	4.8 ± 1.4	4.6 ± 1.2	$P < 0.01$	$P < 0.01$	NS
IBD	4.6 ± 1.0	1.4 ± 1.1	0.2 ± 0.5	$P < 0.01$	$P < 0.01$	$P < 0.01$

Means ± SD (n=22 subjects). SD=standard deviation, NS=not significant, $P > 0.05$, *1 year compared with baseline, †5 year compared with baseline, ‡5 year compared with 1 year

Table 5. Mean clinical and radiographical data at each treated site at baseline and at the 1-and 5-year follow-up periods.

3.4. Conclusion

Compared to PRP with HA, treatment with a combination of cultured periosteal sheets, PRP, and HA led to a significantly more favorable clinical improvement in infrabony periodontal defects at 1-year post-surgery. However, the added benefits of cultured periosteal sheets had statistical and clinical significance. Once a successful clinical treatment has been achieved with a combination of cultured periosteal sheets, PRP, and HA granules, the results are maintainable for 5 years. In addition to the favorable long-term clinical results for treating periodontal osseous defects, the long-term safety of the cultured periosteal sheets in periodontal therapy was established.

4. Application of cultured autogenous periosteal cell sheets to alveolar bone regeneration

Diseases affecting the maxillofacial region often cause defects in alveolar or gnathic bone. The maxillofacial bones are an important determinant of facial features and play a pivotal role in mastication and articulation. Thus, rehabilitation therapy helps morphological and functional recovery of the maxillofacial bones is an important component of the strategies required after the diseases have been treated. Autogenous bone grafting is the most realistic and effective method for regenerating bone tissue in defect sites and is therefore widely used. However, the bone harvesting is accompanied by unavoidable donor site morbidity such as pain and increased risk of infection.

For these types of reasons, we employed the tissue-engineering approach aided by the administration of cells derived from autogenous periosteum, which provided osteoblasts responsible for bone formation and also can promote the recruitment of cells responsible for bone resorption [Nagata *et al.*, 2012, Uematsu *et al.*, 2013a]. Patients scheduled to undergo alveolar bone augmentation prior to dental implant placement were subjected to clinical study of alveolar bone engineering with administration of cultured autogenous periosteal cell (CAPC) sheet.

4.1. CAPCs promoted good bone regeneration and reduced the amounts of bone required for harvesting

CAPCs grafting with autologous bone was performed for alveolar ridge augmentation and for maxillary sinus lift (Fig. 4). The bone was harvested from the anterior region of the mandibular ramus in the majority of cases and from the iliac crest in one case. Bone regeneration after CAPCs grafting with autogenous bone was generally satisfactory, and the predictability of alveolar bone augmentation was high even in large recipient sites. Use of CAPCs resulted in alveolar ridge augmentation with satisfactory morphology and long-term stable bone volume in these advanced cases of atrophy. Furthermore, graft materials containing less than 40% of the autogenous bone particle also demonstrated good bone formation, and it was found that using CAPCs was beneficial for the satisfactory formation and long-term maintenance of regenerated bone. In other words, use of CAPCs can reduce the amount of autogenous bone required, which reduced donor site morbidity during bone harvesting.

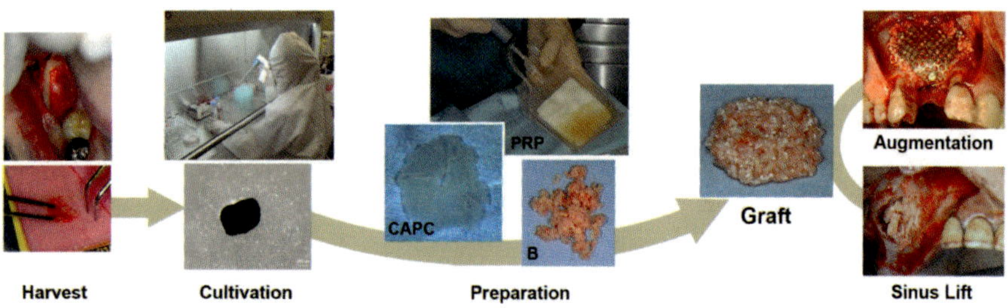

Figure 4. Procedure of autogenous bone grafting with cultured autogenous periosteal cells (CAPCs).

4.2. Recruitment of osteoblasts and osteoclasts to regenerating bone tissue

The histological images of biopsy specimens recovered from the recipient sites at 4 months after CAPCs grafting with particulated bone exhibit newly formed bone in spaces between the cortical bone particles in graft material (Fig. 5). Layers of cells strongly immunopositive for ALP activity were present on the surface of the newly formed bone. Recruitment of TRAP-positive cells was confirmed for grafted autogenous bone particles and the newly formed bone (Fig. 5A–H). In clear contrast to these findings, in the biopsy specimens recovered 4 months after conventional bone grafting, markedly few cells were found around the newly formed bone, and ALP-positive signals, albeit weak, were found exclusively on the surface of the bone (Fig. 5I–L). In addition, recruitment of TRAP-positive cells was negligible. These results indicate that the recruitment of both osteoblasts and osteoclasts to the regenerating bone tissue is markedly activated at the site of CAPCs grafting with autogenous bone particle, as compared with that of conventional bone grafting.

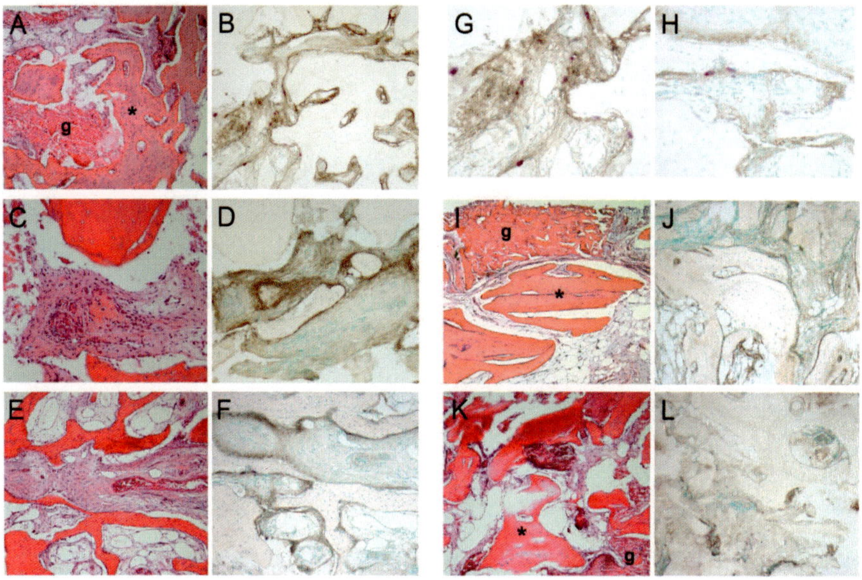

* newly formed bone, g: residual grafted autologous bone particles

Figure 5. Histological analysis of biopsy specimens recovered from the recipient sites at 4 months after bone grafting with CAPCs (A-H) or conventional bone grafting (I-L). A, C, E (H-E staining) and B, D, F, G, H (ALP-TRAP double staining): Active new-bone formation was observed in specimens recovered from the site of a bone graft with CAPCs.

4.3. CAPC-induced bone remodeling after autogenous bone grafting

The volume of newly reformed bone tissue at 1 Y was compared to that at 3 M in the cases of maxillary sinus lift with autologous bone and CAPCs and in the cases of maxillary sinus lift with autologous bone only using reconstructed 3D-CT images. Changes in the mean net volume of augmented bone between the two time points were similar regardless of whether CAPCs were used. When compared after categorizing the area of augmented bone according to CT density, the mean volume was 35% lower at 1 Y than at 3 M in the area with a CT density < 850 HU, regardless of whether CAPCs were used or not (data not shown).

On the other hand, in the high-density area (CT density > 850), marked time-dependent decreases were found only after bone grafting with CAPCs, whereas no decreases were found in conventional bone grafting. In color-coded reconstructed 3D-CT images, high CT density areas believed to correspond to grafted cortical bone particles of mandibular origin contained in the graft material were almost absent in the 3D-CT images taken at 1 Y after autogenous bone grafting with CAPCs, but a smooth surface area with high CT density areas was found on the surface of the augmented bone (Fig. 6). Thus, high-resolution three-dimensional computed tomography (3D-CT) imaging suggested that remodeling of the grafted autogenous cortical bone particles was faster in bone grafting with CAPCs than in conventional bone grafting.

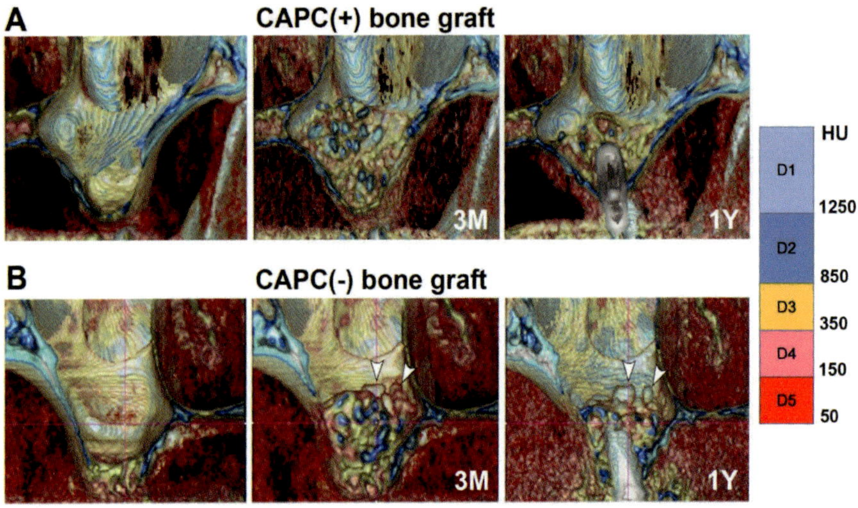

Figure 6. CT images of the maxillary sinus constructed from DICOM data before surgery and at 3 months (3M) and 1 year (1Y) after CAPC(+) bone graft or conventional CAPC(-) bone grafting. Color mapping of the images according to bone density as determined by Hounsfield units (HU) for CT data was applied. In the graft material at 3M, many areas are depicted in light blue (D1) and blue (D2), which is indicative of autogenous cortical bone particles (A and B).

4.4. Conclusions

In this clinical research into CAPC use in bone augmentation, we investigated the effects of CAPCs on bone regeneration in patients. Augmentation that is difficult to achieve by conventional autogenous bone grafting, such as that for building bone in both the directions of height and width in the alveolar ridge or for performing a sinus lift of ≥15 mm in patients with a less than 2-mm-thick maxillary sinus floor, are feasible through the application of CAPCs, regardless of age and sex of the patients. The results of histological and 3D-CT analyses suggest that grafted CAPCs effectively recruited osteoblasts and osteoclasts, thereby simultaneously promoting bone formation and remodeling.

The effects of CAPCs in regenerative bone therapy may be exerted via various biological mechanisms, as illustrated in Figure 7. For instance, CAPCs may serve as the original cell population that gives rise to osteoblasts for bone matrix formation and to constituent cells of bone tissue such as vascular endothelial cells, as well as serving as a source of various growth factors, thereby recruiting and activating osteoclasts.

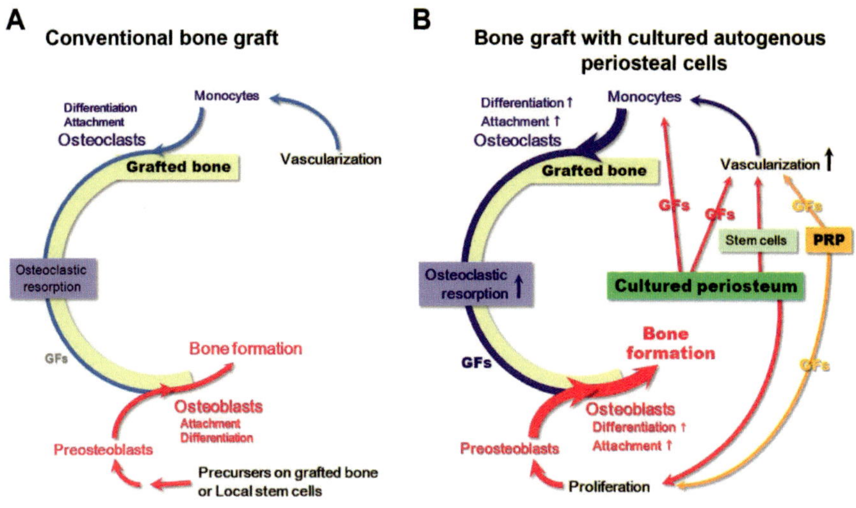

Figure 7. Illustration of hypotheses for bone formation and resorption after (A) conventional bone grafting and (B) after bone grafting with CAPCs.

Bone tissue engineering with CAPC grafting may enable regeneration of bone tissues with complex morphology in a wide area, and thus, the technique will expand the indications for regenerative bone therapy. Currently, the autogenous bone content in the graft material has been reduced to as low as 30% by use of CAPCs. The less invasive approach reduces the burden on patients and also increases the volume of regenerative bone produced. However, autogenous bone particles serve as a secure scaffold for bone augmentation and thus are not replaceable. The ultimate goal is to establish procedures for CAPC culturing and grafting that enable regenerative bone therapy without harvesting bone. To achieve this goal, we must continue working on developing affordable and effective systems for regenerative bone therapy that also satisfy safety requirements.

5. Future perspectives of cell-based therapy for periodontal and bone regeneration

5.1. Progress in regenerative therapy for periodontium and maxillofacial bones

First, material-and scaffold-based therapies have been developed using bone replacement graft and GTR membranes. Second, growth factor application has increased for mixed components and recombinant proteins. Several approaches involving enamel-matrix derivatives and PRP/PRF have been reported, and recombinant growth factors (BMP, FGF-2, PDGF, and GDF-5) have been introduced for periodontal and bone regenerative therapy. Third, cell-based therapy has recently emerged using bone marrow cells, periosteal cells, and periodontal ligament cells. Finally, combination therapy with scaffolds, growth factors, and cells appears to present a promising strategy to achieve the regeneration of large defects and to enlarge the indication of therapy [Egusa *et al.*, 2012]. Figure 8 shows already used materials and possible candidates in the near future for regeneration of periodontium and maxillofacial bones.

5.2. Three challenges for combined cell-based therapy

5.2.1. Biological and clinical safety challenges

Prevention of tumor formation following stem cell implantation is a major safety consideration from the biological point of view. It is critical to have an understanding of the cell genetic instability, culture medium and conditions. In addition, practical errors and accidents involving contamination and infection with virus and mycoplasma should be monitored and avoided.

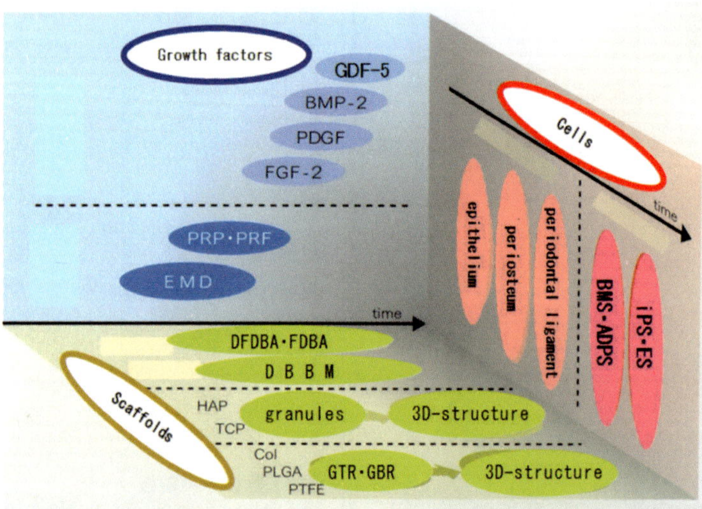

Figure 8. Possible biomaterials, proteins, and cells for regeneration of periodontium and maxillofacial bones.

Growth factor GDF: growth differentiation factor, BMP: bone morphogenetic protein, PDGF: platelet-derived growth factor, FGF: fibroblast growth factor, PRP PRF: platelet-rich plasma platelet-rich fibrin, EMD: enamel matrix derivative

Scaffolds DFDBA: demineralized freeze-dried bone allograft, FDBA: freeze-dried bone allograft, DBBM: demineralized bovine bone matrix, HAP: hydroxyapatite, TCP: tricalcium phosphate, Col: collagen, PLGA: polylactic-polyglycolic acid, PTFE: polytetrafuoroethlene, GTR: guided tissue regeneration, GBR: guided bone regeneration, 3D: three dimensional

Cells BMS: bone marrow stem cells, ADPS: adipose stem cells, iPS: induced pluripotent stem cell, ES: embryonic stem cells

5.2.2. Beneficial technical challenges

It is definitely possible to achieve remarkable regeneration using cell-based therapy compared with conventional regenerative treatment for satisfaction of patients and cost benefit aspects. The matrix and scaffold should have good biocompatibility in terms of the cellular and molecular components during the process of developing and regenerating tissues. The synergistic effects of cell-based therapy with suitable scaffolds and growth factors are anticipated and should be substantiated in the near future.

5.2.3. Challenges of cell delivery system and cell banking

Cell processing center/room (CPC) is required for cell-based therapy, and the refinement of techniques to facilitate laboratory handing of cells is also crucial for biological and clinical applications. CPC should be established at hospital, however, it is impossible to establish CPC. Thus, a well-controlled cell delivery system is necessary for wide area of several prefectures. Furthermore, cell banking may be one strategy to realize the potential of cell-based regenerative therapy. Several types of cells can be cryopreserved to retain regenerative potential. Dental cells can be isolated from the cryopreserved tissue whenever required for future regenerative therapies [Chen et al., 2012].

Author details

Tomoyuki Kawase, Kazuhiro Okuda, Masaki Nagata and Hiromasa Yoshie

Institute of Medicine and Dentistry, Niigata University, Japan

References

[1] Chen FM, Sun HH, Lu H, & Yu Q. (2012) Stem cell-delivery therapeutics for periodontal tissue regeneration. *Biomaterials* 33: 6320-6344.

[2] Egusa H, Sonoyama W, Nishimura M, Atsuta I, & Akiyama K. (2012) Stem cells in dentistry--Part II: Clinical applications. *J Prosthodont Res*56: 229-248.

[3] Engler AJ, Sen S, Sweeney HL, & Discher DE. (2006) Matrix elasticity directs stem cell lineage specification. *Cell* 126: 677-89.

[4] Horimizu M, Kawase T, Tanaka T, Okuda K, Nagata M, Burns DM, & Yoshie H. (2013) Biomechanical evaluation by AFM of cultured human cell-multilayered periosteal sheets. *Micron* 48: 1-10.

[5] Kawase T, Okuda K, Kogami H, Nakayama H, Nagata M, Nakata K, & Yoshie H. (2009) Characterization of human cultured periosteal sheets expressing bone-forming potential: in vitro and in vivo animal studies. *J Tissue Eng Reg Med* 3: 218-229.

[6] Kawase T, Okuda K, Kogami H, Nakayama H, Nagata M, & Yoshie H. (2010a) Osteogenic activity of human periosteal sheets cultured on salmon collagen-coated ePTFE meshes. *J Mater Sci Mater Med* 21: 731-739.

[7] Kawase T, Okuda K, Kogami H, Nakayama H, Nagata M, Sato T, Wolff LF, & Yoshie H. (2010b) Human periosteum-derived cells combined with superporous hydroxyapatite blocks used as an osteogenic bone substitute for periodontal regenerative therapy: An animal implantation study using nude mice. *J Periodontol* 81: 420-427.

[8] Kawase T, Tanaka T, Nishimoto T, Okuda K, Nagata M, Burns DM, & Yoshie H. (2011) Improved adhesion of human cultured periosteal sheets to a porous poly(L-lactic acid) membrane scaffold without the aid of exogenous adhesion biomolecules. *J Biomed Mater Res A* 98: 100-113.

[9] Kawase T, Uematsu K, Kamiya M, Nagata M, Okuda K, Burns DM, Nakata K, & Yoshie H. (2014) Real-time quantitative polymerase chain reaction and flow cytometric analyses of cell adhesion molecules expressed in human cell-multilayered periosteal sheets in vitro. *Cytotherapy* 16: 653-661.

[10] Nagata M, Hoshina H, Li M, Arasawa M, Uematsu K, Ogawa S, Yamada K, Kawase T, Suzuki K, Ogose A, Fuse I, Okuda K, Uoshima K, Nakata K, Yoshie H, Takagi R. (2012) A clinical study of alveolar bone tissue engineering with cultured autogenous periosteal cells: coordinated activation of bone formation and resorption. Bone 50: 1123-1129.

[11] Okuda K, Kawase T, Momose M, Murata M, Saito Y, Suzuki H, Wolff LF, & Yoshie H. (2003) Platelet-rich plasma contains high levels of platelet-derived growth factor and transforming growth factor-ß and modulates the proliferation of periodontally related cells in vitro. *J Periodontol* 74: 849-857.

[12] Okuda K, Tai H, Tanabe K, Suzuki H, Sato T, Kawase T, Saito Y, Wolff LF, & Yoshie H. (2005) Platelet-rich plasma combined with a porous hydroxyapatite graft for the treatment of intrabony periodontal defects in humans: a comparative controlled clinical study. *J Periodontol* 76: 890-898.

[13] Okuda K, Kawase T, Nagata M, Yamamiya K, Nakata K, Wolff LF, & Yoshie H. (2009) Treatment of human infrabony periodontal defects by grafting human cultured periosteum sheets combined with platelet-rich plasma and porous hydroxyapatite granules: case series. *J Int Acad Periodontol* 11: 206-213.

[14] Okuda, K. Kawase, T. Nagata, M. Yamamiya, K. Nakata, K. Wolff, LF. & Yoshie, H. (2013) Tissue engineered cultured periosteum sheet application to treat infrabony defects: case series and 5-year results. *Int J Periodont Rest Dent* 33: 281-287.

[15] Uematsu K, Kawase T, Nagata M, Suzuki K, Okuda K, Yoshie H, Burns DM, & Takagi R. (2013a) Tissue culture of human alveolar periosteal sheets using a stem-cell culture medium (MesenPRO-RS™): In vitro expansion of CD146-positive cells and concomitant upregulation of osteogenic potential in vivo. *Stem Cell Res* 10: 1-19.

[16] Uematsu K, Nagata M, Kawase T, Suzuki K, & Takagi R. (2013b) Application of stem cell media to explant culture of human periosteum: an optimal approach for preparing osteogenic cell material. *J Tissue Eng* in press.

[17] Vincent LG, Choi YS, Alonso-Latorre B, del Álamo JC, & Engler AJ. (2013) Mesenchymal stem cell durotaxis depends on substrate stiffness gradient strength. *Biotechnol J* 8: 472-84.

[18] Yamamiya K, Okuda K, Kawase T, Hata K, Wolff LF, & Yoshie H. (2008) Tissue-engineered cultured periosteum used with platelet-rich plasma and hydroxyapatite in treating human osseous defects. J Periodontol 79: 811-818.

Chapter 3

Pulp Regeneration by Harnessing Dental Pulp Stem Cells

Misako Nakashima and Koichiro Iohara

Additional information is available at the end of the chapter

http://dx.doi.org/10.5772/58905

1. Introduction

Teeth possess multiple functions. Besides mastication, they are important for taste, tactile sense, speech and esthetics. The longevity of teeth is critical for the maintenance and enhancement of quality of life. Over fifty percent of extracted teeth are caused by caries and root fracture. Deep caries and pulp exposure have been treated by pulp capping or partial pulp amputation to preserve pulp tissue, with limited success. In case of irreversible pulpitis,the entire pulp has to be removed and the root canal space has to be completely filled with artificial materials after disinfection. However, incomplete disinfection and coronal leakage cause periapical disease. The endodontic success rate is less than 50-70% for retreatment of those endodontically failed teeth with periapical lesions and/or some clinical symptoms. Enlargement and debridement of infected root canals render teeth brittle. The post-operative procedures with post preparation also increase the possibility of root fracture. Therefore, loss of pulp vitality is highly associated with root fracture and tooth loss. There has been no superior synthetic material to replace natural pulp and dentin. Thus, regenerating pulp-dentin tissues seems to be an ideal approach to restore tooth functions compromised by pulp injury and/or inflammation (Nakashima and Akamine, 2005; Nakashima, et al., 2009). Under natural conditions, pulp and dentin have limited capacity to regenerate and repair when damaged. This chapter describes recent advances and promising approaches for pulp regeneration, including isolation and characteristics of pulp stem/progenitor cells and partial and complete regeneration of pulp. The complete pulp regeneration with other tissue stem cells is also addressed.

2. Pulp regeneration in mature teeth with complete apical closure

2.1. Pulp stem/progenitor cells suitable for pulp regeneration

Dental pulp tissue is rich in vasculature and innervation. There is an intimate association of innervation with vasculature of the dental pulp, involving in pulp homeostasis. Thus, angiogenesis and re-innervation are critical prerequisites for pulp regeneration (Nakashima and Akamine, 2005). Numerous studies have reported the isolation of adult pulp stem/ progenitor cells with high proliferative and multi-differentiation potential in vitro. However, only a few studies have demonstrated angiogenesis and re-innervation in vivo after transplantation.

Porcine and human pulp $CD31^-$side population(SP) cells and CD105/endoglin$^+$cells, positive for CD29, CD44, CD73, and Thy-1/CD90, and negative for CD31, CD45 and CD133, increase the blood flow including high density of capillary formation after transplantation into mouse hindlimb ischemia (Iohara et al., 2008; Nakashima et al., 2009). SP cells are enriched in the 'true' or 'mother' adult stem cells, which exhibite lower level of the DNA binding fluorescent dye, Hoechst 33342than the rest of pulp cells. CD105/endoglin is a component of the transforming growth factor-beta receptor complex. CD105 is widely expressed on mesenchymal stem cells (MSCs), and the positive isolation strategy based on the selection of the cells that express CD105 has been proposed for isolation of MSCs. Pulp $CD31^-$cells and $CD105^+$cells induce angiogenesis and neurogenesis in rat cerebral ischemia by releasing neurotrophic factors, such as VEGF in peri-infarct area and promoting migration and differentiation of the endogenous neuronal progenitor cells, and accelerate the functional recovery (Sugiyama, et al., 2011). Local transplantation of these pulp stem/progenitor cells results in successful engraftment in proximity to the newly formed vasculature without direct incorporation into vessels, suggesting enhancement of angiogenesis and re-innervation by trophic effects (Iohara, 2008; Sugiyama, 2011). The conditioned medium of pulp stem/progenitor cells contain high concentration of angiogenic/neurotrophic factors such as brain-derived neurotrophic factor (BDNF), nerve growth factor (NGF), vascular endothelial growth factor (VEGF-A), glial cell-line derived neurotrophic factor (GDNF), granulocyte-macrophage colony-stimulating factor (GM-CSF), granulocyte colony-stimulating factor (G-CSF), matrix metalloproteinase 3 (MMP3). These factors have effect on proliferation, migration, and anti-apoptosis (Iohara, 2008, Sugiyama, 2011). On the other hand, unfractionated total pulp cells express lower level of angiogenic/ neurotrophic factors, and are less effective on angiogenesis and neurogenesis compared to $CD31^-$SP cells and $CD105^+$cells after transplantation in hindlimb ischemia and brain ischemia (Nakashima, et al., 2009). Thus, potential utility of some subfractions of pulp stem/progenitor cells has been demonstrated as cell sources for cell-based therapy of angiogenesis, re-innervation and pulp regeneration. MSCs secrete soluble cytokines and growth factors that function in a paracrine fashion, contributing to repair and regeneration by enhanced cell survival, angiogenesis, neurogenesis and neuroprotection, and by activating endogenous stem cells (Kassis et al., 2011).

2.2. Partial pulp regeneration with pulp stem/progenitor cells

Recent advances in tissue regeneration have led to the possibility of partial pulp regeneration in case of pulp exposure or pulpotomy of partial pulptitis. There are two methods in partial pulp regeneration even in mature teeth with complete apical closure; (i)transplantation of pulp stem/progenitor cells with scaffold or engineered pulp tissue, and (ii) application of migration/homing factors with scaffolds.

The first method (Figure 1A) is evidenced by an experiment in which canine autologous pulp $CD31^{-}SP$ cells are transplanted in the cavity on the amputated pulp one day after three dimensional cultures with collagen scaffold. The cavity on the amputated pulp is filled with the pulp tissue with well-developed vasculature and innervations (Figure 2A, C-E). The tubular dentin is formed along the dentinal wall (Figure 2F, G) (Iohara, et al., 2009). Many of transplanted $CD31^{-}SP$ cells are in the proximity of the newly formed capillaries and express angiogenic/neurotrophic factors, implicating trophic effects on neovascularization (Iohara et al., 2009). On the other hand, after transplantation of $CD31^{+}SP$ cells, fewer capillaries and lower volume of the regenerated tissue are seen in the cavity on the amputated pulp compared with transplantation of $CD31^{-}SP$ cells (Figure 2C)(Iohara, et al., 2009), suggesting different potential utility in induction of pulp regeneration among subfractions of pulp stem/progenitor cells.

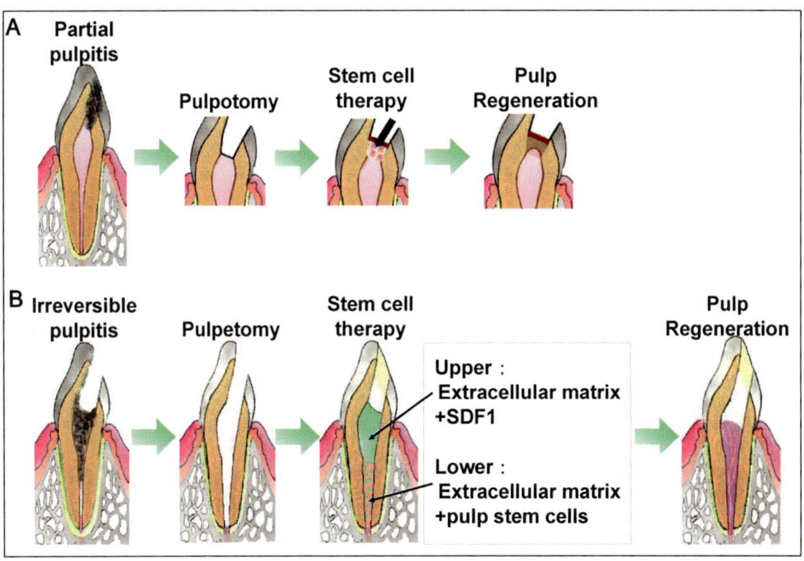

Figure 1. Schematic diagrams of a canine model for partial and complete pulp regeneration in permanent mature teeth. (A) Partial pulp regeneration model. (B) Complete pulp regeneration model. (Iohara et al., Tissue Eng. 2011)

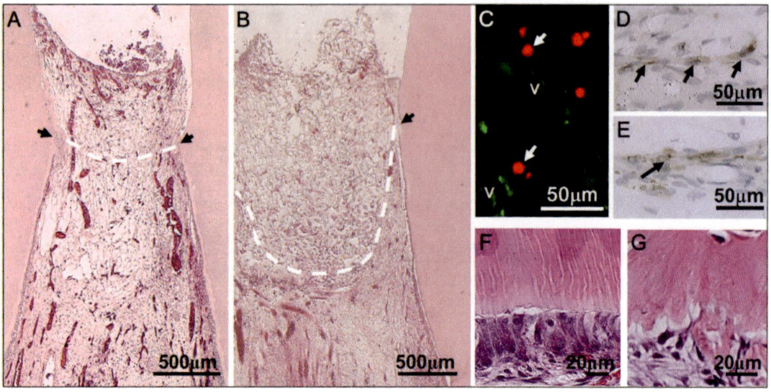

Figure 2. Neovascularization and pulp regeneration in the cavity on the amputated pulp of dogs after autogenous transplantation of CD31$^-$/CD146$^-$SP cells. (A, B) Fourteen days after transplantation of a three-dimensional pellet on day 1 of culture with type I and type III collagen. H.-E. staining. Amputated sites of dental pulp are indicated by arrows. The established pulp tissue is marked by dotted lines. (A) CD31$^-$/CD146$^-$SP cells. The cavity in the amputated pulp is filled with regenerated pulp tissue. Capillaries extended to the top of the tissue beneath the site of the filling cement. (B) CD31$^+$/CD146$^-$SP cells. Few capillaries are seen in the cavity on the amputated pulp despite engraftment of the transplanted pellet. (C) Immunostaining of CD146. DiI-labeled CD31$^-$/CD146$^-$SP cells (arrows) can be seen close to the newly formed capillaries (V). (D, E) Immunostaining of neurofilament (arrows).(D) in the regenerated pulp tissue on the amputated site. (E) in the normal pulp tissue under the amputated site. (F) Tubular dentin formation along the dentinal wall in the cavity (G) Osteodentin formation at the top of the cavity under cement. (Iohara et al., Reg Med. 2009)

2.3. Complete pulp regeneration with pulp stem/progenitor cells and migration factors

Complete pulp regeneration is optimistically considered by transplantation of autologous pulp stem/progenitor cells in case of irreversible pulpitis of even mature teeth with complete apical closure. The transplanted cells should survive with nutrition and oxygen in the pulpectomized root canals before revascularization. Some subfractions of pulp stem/progenitor cells, such as CD105$^+$cells or CD31$^-$SP cells, have more regenerative potential compared with unfractionated total pulp cells. Pulp stem/progenitor cells are capable to induce angiogenesis and re-innervation without direct incorporation into vasculature. Thus, the induction of MSC recruitment from surrounding tissues or from the circulation by migration/homing factors together with transplanted stem/progenitor cells can be a helpful modality to initiate and support cell therapy for pulp regeneration. Pulp CD31$^-$SP cells and CD105$^+$cells are CXCR4-positive, having high migratory and proliferative activities with SDF-1 (Iohara, et al., 2008; Nakashima, et al., 2009). Thus, one candidate for migration/homing factors is SDF-1, a chemokine for CXCR4-positive stem cells. It is plausible that SDF-1 transplanted with pulp stem cells provides signals for homing and proliferation of endogenous stem cells originating from surrounding periodontal ligament, bone marrow and vasculature.

Optimal scaffolds should also be developed for efficient and safe complete pulp regeneration. It is critical for the scaffold to be injected into the root canal with optimal flow, devoid of bubble formation and mechanical properties after hardening in vivo. In addition the scaffold has to be biocompatible, with high bioactivity to integrate trophic factors secreted by stem cells. Furthermore it should be biodegradable by releasing these factors to replace the damaged tissue by newly regenerated tissue without immune response. In addition, the scaffold should not stimulate mineralization and differentiation of odontoblasts and/or osteodentinoblasts in the root canal except along the dentinal wall for pulp regeneration (Nakashima, et al., 2009). Potential scaffolds include natural polymers, such as collagen and gelatin with good biocompatibility and bioactivity (Nakashima and Akamine, 2005).

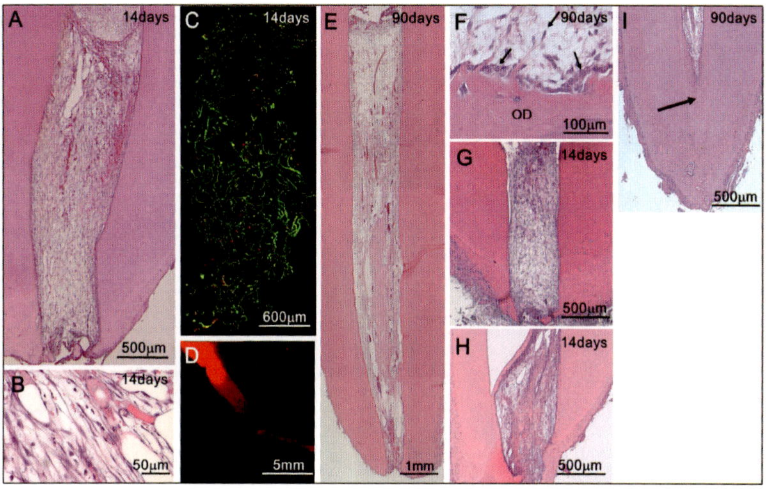

Figure 3. Complete pulp regeneration after autologous transplantation of CD105⁺cells with SDF-1 in the pulpectomized root canal in dogs. (A-F) CD105⁺cells with SDF-1. (C) Three dimensional images of new vascularization by whole mount immunostaining with lectin. (D) DiI-labeling from the upper part of the regenerated pulp on day 14. Note inferior alveolar nerve connecting to the regenerated pulp, suggesting re-innervation. (F) Odontoblastic cells (arrows) lining to newly formed osteodentin/tubular dentin (OD) along with the dentinal wall. (G) CD105⁺cells only. (H) SDF-1 only. (I) Total pulp cells with SDF-1. Mineralized tissue (arrow) and osteodentin. (Iohara et al., Tissue Eng. 2011)

Thus, based on the concepts described above, pulp regeneration has been demonstrated by using the triad, stem/progenitor cells, migration/homing factors and optimal scaffolds (Nakashima, et al., 2011, Iohara, et al., 2011)(Figure 1B). Pulpectomy with enlarged apical portion, 0.6~0.7 mm in width, was performed in mature teeth with complete apical closure in an experimental model in dogs (Iohara, et al., 2011, Ishizaka et al., 2012). Current evidence shows that the root canal is filled with pulp-like loose connective tissue with vasculature and nerves after autologous transplantation of pulp stem/progenitor cells, $CD31^{-}SP$ cells or pulp $CD105^{+}$cells, 5×10^{5} cells in cell number, with SDF-1 in collagen type I and type III scaffold (Figure 3A-F) (Nakashima, et al., 2011, Iohara, et al., 2011, Ishizaka et al., 2012). The three dimensional image of induced vascularization in the regenerated tissue (Figure 3C) is similar in density and orientation to those in the normal pulp. The neuronal process from regenerated pulp is connecting to the inferior alveolar nerve (Figure 3D)(Iohara et al., 2011). The transplanted pulp stem cells expressed angiogenic/neurotrophic factors, and localized in the vicinity of newly formed capillaries (Iohara et al., 2011; Ishizaka et al., 2012), suggesting potential trophic effects on angiogenesis. The odontoblast-like cells attached to the dentinal wall in the root canal, and produce dentin-like mineralized tissue extending their processes into dentinal tubules (Figure 3F). The enlarged apical portion following pulpectomy is filled by additional formation of dentin and cementum. On the other hand, transplantation of stem/progenitor cells alone (Figure 3G), or SDF-1 alone (Figure 3H), yield significantly less pulp tissue. When unfractionated total pulp cells are implanted in place for the fractionated pulp stem cells, the regenerated tissue is significantly less in volume and undergo mineralization (Figure 3I) (Iohara et al., 2011). The regenerated tissue induced by the pulp stem/progenitor cells and SDF-1 is identical to be normal functional pulp tissue as demonstrated similar expression of the pulp tissue markers, *Syndecan* and *TRH-DE* mRNA. It is noteworthy that the qualitative and quantitative protein and mRNA expression patterns obtained by two dimensional electrophoretic analyses and microarray analyses are virtually identical (Iohara et al., 2011; Ishizaka et al., 2012). Furthermore, in a canine experimental model of periapical disease, which root canal is kept open for more than three weeks, pulp regeneration has been demonstrated after transplantation of pulp stem cells and SDF-1 as demonstrated in case of pulpitis. The possible mechanisms of pulp regeneration may involve the CXCR4/SDF-1 axis functioning as a migration/homing factor for CXCR4-positive endogenous stem cells to migrate to the coronal portion of the root canal, proliferate, and differentiate into endothelial cells for angiogenesis and re-innervation by angiogenic and neurotrophic factors secreted by transplanted stem cells. These findings suggest potential clinical translation of complete pulp regeneration by harnessing pulp stem/progenitor cells with high angiogenic/neurogenic potential and additional migration/homing factors in endodontic treatment. The mechanisms of recruitment and crucial molecules for cell migration are still unclear. Chemokines and their receptors which play critical roles need further scrutiny.

2.4. Other sources of tissue stem cells for pulp regeneration

Supply of autologous pulp tissue declines with age, and alternative sources of MSCs to pulp stem cells are necessary for clinical application in endodontic treatment. Some transcriptional and epigenetic analyses have revealed very similar profiles among a variety of MSCs from bone marrow, adipose tissue, placenta, umbilical cord and amnion (Aranda etal., 2009; Boeuf and Richter, 2010). Different expression profile of trophic factors and growth factors, however, has also been reported among MSC populations (Noël et al., 2008; Boeuf and Richter, 2010; Philippe et al., 2010). Thus, it is necessary to investigate the requirements and preconditions of MSCs for effective induction of pulp regeneration (Ishizaka, et al., 2012). A candidate of an alternative cell source for pulp regeneration is autologous bone marrow and adipose tissue-derived MSCs due to neither ethical nor immunoreactive considerations. Transplantation of adipose CD31$^-$SP cells resulted in formation of a similar amount of regenerated tissue compared with that of pulp CD31$^-$SP cells (Figure 4A, C). Bone marrow CD31$^-$SP cell transplantation induce significantly less amount of regenerated tissue compared with other two (Figure 4B, D). Those regenerated tissues, however, are all identical to pulp tissue confirmed by morphological and functional similarity, mRNA expression patterns of microarray analyses and two-dimensional electrophoretic analyses (Ishizaka, 2012). Thus, it is suggested that MSCs from different sources are comparable in pulp regeneration in vivo and that their capabilities are not influenced by the niche of their origin. Enhanced matrix formation in adipose cell transplantation, which lead to root canal obliteration after longer period, suggest that some trophic factors, growth/differentiation factors and/or inhibitors of stemness may influence on active pathway for differentiation in adipose stem cells. Further studies are needed to elucidate its involved pathways and processes.

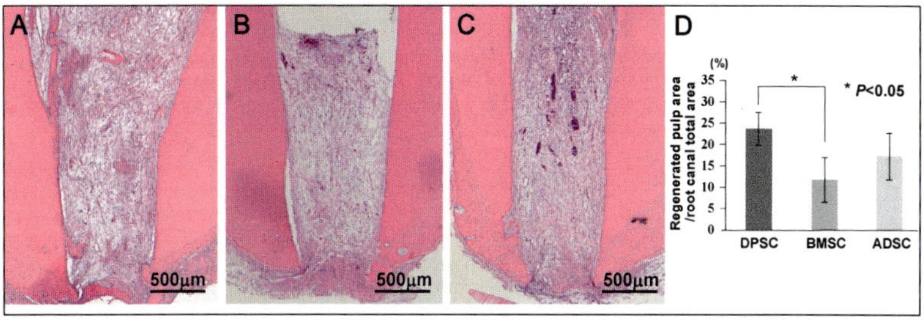

Figure 4. Complete regeneration of pulp tissue after autologous transplantation of $CD31^-SP$ cells with SDF-1 in the emptied root canal after pulpectomy in dogs on 14days. (A) pulp $CD31^-SP$ cell transplantation. (B) bone marrow $CD31^-SP$ cell transplantation. (C) adipose $CD31^-SP$ cell transplantation. (D) Ratio of regenerated area to root canal area. Data are expressed as means ± SD of 5 determinations. (Ishizaka et al., Biomaterials. 2012)

3. Conclusion

Cell therapy with subfractions of pulp stem/progenitor cells, $CD31^-SP$ cells or pulp $CD105^+$ cells is capable to induce pulp regeneration, presenting the following features: i) vascularization, ii) similar cell density and architecture of the extracellular matrix as the natural pulp, iii) new odontoblasts aligning with the dentinal wall to produce new dentin, and iv) innervation.

$CD31^-SP$ cells have to be labeled with DNA binding Hoechst 33342 and isolated by flowcytometry, which requires the confirmation of safety. $CD105^+$cells have to be isolated by magnetic antibody beads method if not by flowcytometry, and have to be cost effective. The unfractionated total pulp cells by colony formation method are less suitable for pulp regeneration since the regenerated tissue is much less in volume and much more to undergo mineralized compared with fractionated stem/progenitor cells. A novel isolation method capable of confirmation of safety and efficiency from a small amount of pulp tissue should be developed prior to clinical trials. Although there are still some hurdles to overcome in regenerative endodontics, there are the rewards of immense clinical advantages and benefits for patients to maintain the function and longevity of teeth resulting in maintenance of whole body and enhancement of quality of life.

Acknowledgements

This work was supported by Collaborative Development of Innovative Seeds, Potentiality Verification Stage from Japan Science and Technology Agency to MN; and the Research Grant for Longevity Sciences (21A-7) from the Ministry of Health, Labour and Welfare to MN.

Author details

Misako Nakashima and Koichiro Iohara

National Center for Geriatrics and Gerontology, Research Institute, Japan

References

[1] Aranda, P.; Agirre, X.; Ballestar, E.; Andreu, E.J.; Román-Gómez, J. & Prieto, I. et al. (2009). Epigenetic signatures associated with different levels of differentiation potential in human stem cells. *PLoS One* 4(11):e7809.

[2] Boeuf, S. & Richter, W. (2010). Chondrogenesis of mesenchymal stem cells: role of tissue source and inducing factors. *Stem Cell Res Ther* 1(4):31.

[3]

[4] Iohara, K.; Zheng, L.; Ito, M.; Ishizaka,R.; Nakamura, H. & Into T. et al. (2009). Regeneration of dental pulp after pulpotomy by transplantation of $CD31^{-}/CD146^{-}$side population cells from a canine tooth. *Regen Med* 4(3): 377-385.

[5] Iohara, K.; Imabayashi, K.; Ishizaka, R.; Watanabe, A.; Nabekura, J.; Ito, M.; Matsushita, K.; Nakamura, H. & Nakashima, M. (2011). Complete pulp regeneration after pulpectomy by transplantation of CD105+ stem cells with stromal cell-derived factor-1. *Tissue Eng Part A* 17(15-16): 1911-1920.

[6] Ishizaka, R.; Iohara, K.; Murakami, M.; Fukuta, O. &Nakashima, M. (2012). Regeneration of dental pulp following pulpectomy by fractionated stem/progenitor cells from bone marrow and adipose tissue. *Biomaterials* 33(7):2109-2118.

[7] Nakashima, M. &Akamine, A. (2005). The application of tissue engineering to regeneration of pulp and dentin in endodontics. *J Endod* 31(10):711-718.

[8] Kassis, I.; Vaknin-Dembinsky, A. & Karussis, D. (2011). Bone marrow mesenchymal stem cells: agents of immunomodulation and neuroprotection. *Curr Stem Cell Res Ther* 6(1):63-68.

[9] Nakashima, M.; Iohara, K. & Sugiyama M (2009). Human dental pulp stem cells with highly angiogenic and neurogenic potential for possible use in pulp regeneration. *Cytokine & Growth Factor Reviews* 20(5-6):435-440.

[10] Nakashima, M. & Iohara, K. (2011). Regeneration of dental pulp by stem cells. *Adv Dent Res* 23(3): 313-319.

[11] Noël, D.; Caton, D.; Roche, S.; Bony, C.; Lehmann, S. & Casteilla L., et al. (2008). Cell specific differences between human adipose-derived and mesenchymal-stromal cells despite similar differentiation potentials. *Exp Cell Res* 314(7):1575-1584.

[12] Philippe, B.; Luc, S.; Valérie, P.B.; Jérôme, R.; Alessandra, B.R. & Louis, C.(2010). Culture and use of mesenchymal stromal cells in phase I and II clinical trials. *Stem Cells Int* 2010:503593.

[13] Sugiyama, M.; Iohara, K.; Wakita, H.; Hattori, H.; Ueda, M.; Matsushita, K. & Nakashima, M. (2011). Dental pulp-derived CD31(-)/CD146(-) side population stem/progenitor cells enhance recovery of focal cerebral ischemia in rats. *Tissue Eng Part A* 17(9-10):1303-1311.

Chapter 4

Regenerative Medicine in the Oral and Maxillofacial Region

Hideto Saijo

Additional information is available at the end of the chapter

http://dx.doi.org/10.5772/58903

1. Introduction

The bone reconstruction has conventionally used autologous bone, allograft bone, or artificial bone in the maxillofacial resigns. Since bone reconstruction using artificial bone avoids the highly invasive harvesting of bone needed for autologous bone grafts, many different types of artificial bone have been developed. The requirements of artificial bone are that it (1) has good surgical manipulability, (2) can conform well to the defective area, (3) has moderate strength, and (4) has the capacity to induce bone regeneration. Artificial bone is marketed worldwide as blocks, in particulate form, or as paste, but no artificial bone has yet been developed that satisfies all these requirements. The ability to reproduce shapes is particularly important, and from the point of view of esthetics, it is regarded as the key to success in the field of cranio-maxillo-facial reconstruction. Reconstruction using custom-made artificial bone and titanium mesh trays has been reported, (Iino M et al., 2009). but since the reconstructions were made using computer-aided design, it would be difficult to customize the reconstruction to the level of detail that surgical operators expect. The clinical application of custom-made artificial bone that we have developed is described.

2. Materials and methods

The subjects were patients with a defect of the maxilla or mandible that was either congenital or acquired through trauma. Cranial plain computed tomography (CT) imaging was carried out according to the usual acquisition protocols using the Aquilion 16 scanner (Toshiba Medical Systems Corporation, Japan) at 120 kV, 300 mA, 0.5-mm slice interval, and helical pitch of 16. CT images were saved in Digital Imaging and Communications in Medicine (DICOM) format for transmission over a network. CT values were determined for extracting bony tissue from the slice images, and the acquired 3-dimensional data were output as Standard Template Library (STL) files. The 3-dimensional form of the skeletal structure to be fabricated was designed using 3-D computer-aided design (CAD) on the basis of the STL data, and a solid model was fabricated using the Z406 inkjet powder layering device.

2.1. Surgical simulations and artificial bone fabrication

The surgical simulations were carried out by sculpting an original wax into defective areas or areas needing supplements of artificial bone on the solid models (Fig. 1). Since the response of soft tissues to changes differs depending on the individual, the wax needed to be shaped with the condition of the soft tissues taken into account. The simulation solid models were scanned by CT, and the CT data were output in DICOM format. Using the CAD data, the regions of artificial bone sculpted in wax were extracted (Fig. 2). Using the CAD system, the inner surface of the artificial bone structure that makes contact with the host bone and the holes for fixing the artificial bone in place were designed onto the extracted artificial bone data. The artificial bone was then fabricated using the Z406 3D printer (DICO), an inkjet powder layering device. The material for fabricating the artificial bone was α-tricalcium phosphate powder, with aqueous solution of polysaccharides as a sclerosing solution. The artificial bone thus formed was a calcium phosphate. (Saijo H et al., 2011).

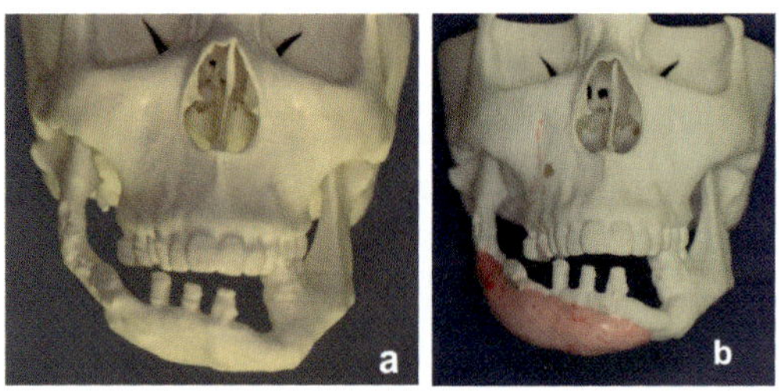

Figure 1. Simulation with wax. a: The three-dimensional model that was molded with gypsum. b: Simulation with the wax.

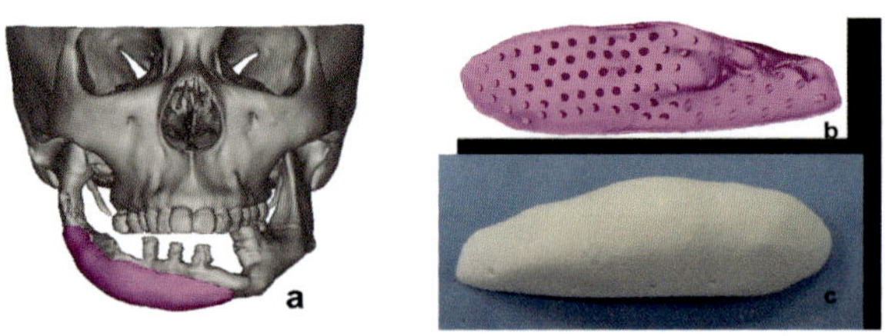

Figure 2. The design of artificial bone. a: The artificial bone designed by CAD. b: The inside view of the artificial bone in the CAD. The inside structure of the artificial bone can be freely simulated by CAD. c: The artificial bone that was made.

2.2. Surgical procedures and follow-up

The operation was performed under general anesthetic, and the region was approached both from the mouth and externally. The periosteum was detached with care not to damage it, and the bone of the graft site was revealed. A trial insertion was made of the custom-made artificial bone, and changes in the soft tissues were checked. Where fixation was then carried out, holes for fixing were made using a surgical drill, and the artificial bone was fixed in place with absorbable polyglycolic suture in several places. The surgical incision was closed by covering the artificial bone with periosteum and making periosteal sutures in order to prevent any deviation resulting from movement of the artificial bone after surgery(Saijo H et al., 2008).

3. Results

Conventional autologous grafts take several hours because the autologous bone must be harvested and shaped, but the present procedure greatly reduced the operation time. During surgery, there was excellent conformity between the artificial bone and the host bone, and firm fixing was not needed. Even where comparatively large artificial bone was grafted and fixing was necessary, no problems were encountered with fixing in several places. Furthermore, at around 3 months after surgery, there were appearances of partial bone union between the artificial bones and the host bone tissue in some patients. In addition, the patients showed a high satisfaction with the facial features following surgery.

3.1. Case 1 (Fig.3)

The patient was 55 years old woman. The patient visited the clinic with a main complaint of facial asymmetry. She was treated for cancer on the left side of her tongue approximately 20 years previously with partial resection of the tongue and segmental resection of the mandible. The region was reconstructed with grafts of a rib and a pectoralis major musculo-cutaneous flap. Radiation therapy was subsequently administered as after-treatment. A graft from the left ilium was carried out 5 years later due to resorption of the first graft. Implants were subsequently placed on the maxilla and the mandible, and dental implants were fitted, but noticeable facial asymmetry remained.

Mandibular reconstructive surgery consisting of an artificial bone graft was carried out under general anesthetic. The approach was made from outside the mouth, as there was a scar present on the neck from the previous operation. The bone was detached by subperiosteal dissection and the host bone graft site revealed, and a trial insertion of the custom-made artificial bone was made. Improvement of the facial features was confirmed, and the artificial bone was fixed to the host bone with 2-0 Vicryl suture. After placement of the graft, the artificial bone was covered with part of the periosteum and the fatty layer, and the surgical wound was closed in the usual way. Presently, at over 1 year operation, the patient is progressing and there is improvement of the facial features. No inflammatory reaction or other abnormalities have been found, and CT images show the position of the artificial bone to be stable with partial bone union.

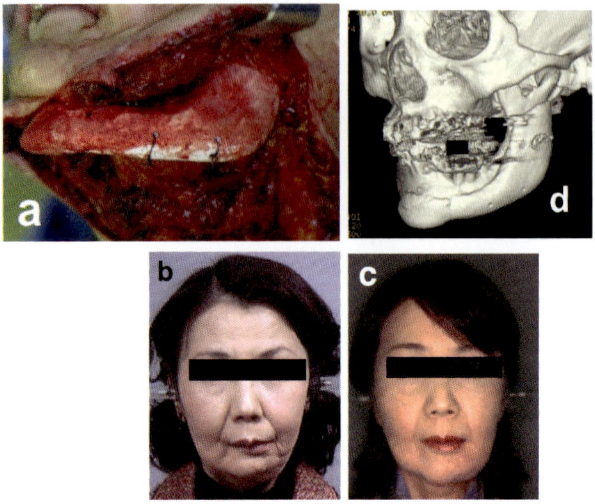

Figure 3. Case 1 a: During operation. b: Facial appearance of Patient 2 before surgery. c: Facial appearance at 2 year after surgery. d: 3D CT image of the left lower jaw after surgery

3.2. Case 2 (Fig.4)

The patient was 31 years old woman. The patient was diagnosed with sarcoma of the mandible when she was 20 years old, and she was treated with segmental resection of the mandible and reconstruction. Mandibular reconstructive surgery was subsequently carried out on several occasions using donor bone from the rib and fibula, but she visited this department in 2001 due to bone resorption resulting from infection. Sequestrum removal and reconstruction using a vascularized fibular osteocutaneous flap were carried out at our department, but since there was still partial resorption due to infection, a graft of bone from a rib was carried out. The engrafted bone was subsequently stable, and dental implants were placed and the denture fitted. However, the facial features changed markedly as a result of the repeated infections and bone grafts, and an artificial bone graft was planned for the present treatment. The incision line from the previous surgery was used for the approach. The graft site bone was revealed, and a trial insertion of the custom-made artificial bone was made. Improvement of the facial features was confirmed, and the artificial bone was fixed to the host bone with several stitches of absorbable sutures. At 6 months after surgery, no abnormalities were found and there was improvement of the facial features. In addition, there were no abnormalities in the position of the artificial bone.

http://dx.doi.org/10.5772/58903

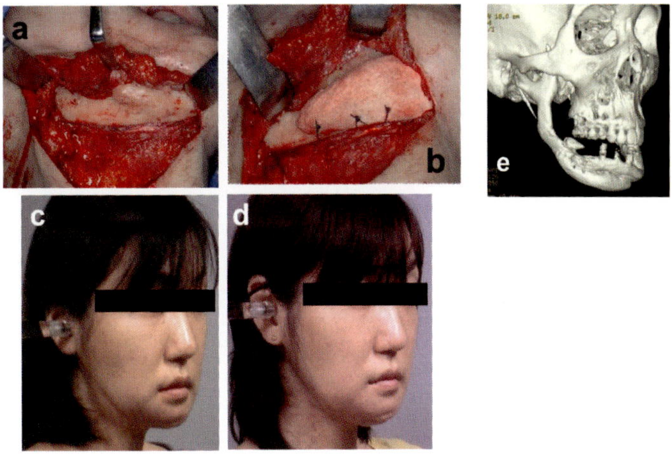

Figure 4. Case2 a, b: During operation c: Facial appearance before surgery d: Facial appearance at 1 year after surgery. e: 3D CT image of the left lower jaw after surgery

4. Discussion

Reconstruction with hard tissue is extremely useful in the treatment of bone defects in the cranio-maxillo-facial region. However, since reconstruction using autologous bone requires harvesting of the bone, massively invasive intervention is unavoidable. Particularly in cases of a large bone defect, it is not only necessary to use a revascularized osteocutaneous flap, which requires the microvessels to be anastomosed, but also craftsmanship is needed to shape the bone to conform to the graft site. Allograft bone carries the risk of infection, and there are also ethical issues, in addition to which the bone needs to be crafted into shape to conform to the graft site. The advantages of artificial bone are that no harvesting site is needed, it has excellent biocompatibility, and the surgical technique is brief and uncomplicated. The artificial bones currently being used around the world can be divided into hydroxyapatite15), α-TCP (α-tricalcium phosphate), and α-TCP (α-tricalcium phosphate). Each has its own characteristics, and the clinical application depends on the nature of the region of defective bone. Hydroxyapatite (HA) artificial bone has a strong affinity for biological tissue and is characterized by direct binding to bony tissue. This is because new bone growth is induced even inside pores by the invasion of collagen tissue, which is needed for ossification. In addition, HA has better mechanical strength than human bone, and it is widely used in clinical practice as a material for replenishing bone defects. (Saijo H et al., 2010). It has been made into commercially available products as porous blocks, granules, and paste. However, HA is highly crystalline and shows little resorption, almost all of it remaining in a stable condition. Thus, HA is processed at high temperature and is highly crystalline, and it is a biomaterial that is not readily resorbed. Tricalcium phosphate(TCP), on the other hand, is resorbed and replaced as the bone is restored. It has excellent tissue affinity and is osteoconductive. In other words, TCP is gradually converted to HA in vivo, and there is a mechanism whereby, as resorption of the artificial bone progresses, it is simultaneously replaced with new bone that forms as a result of invasion by osteoblasts. Thus, the artificial bone, which has excellent osteoconductivity and biocompatibility, is ultimately replaced by bone. In the present study, the main component of artificial bone was fine α-TCP powder, which was made into a non-sintered hardened body of TCP through the use of a sclerosing solution. Such hardened bodies of TCP are likely to be biochemical precursors of HA, and not only do they have excellent biocompatibility like HA, but they are also artificial bones that are resorbed and replaced in vivo. It therefore appears likely that the artificial bones grafted in the present study will be replaced by bone.

Another feature of the present method is that the artificial bone is fabricated from a wax-up made by the surgical operator prior to surgery, using the 3-D cast. This means that the present method is able to reproduce the required shape more faithfully than conventional methods of fabrication. This results in better compatibility than existing types of custom-made artificial bone, so that practically no adjustments are necessary during the surgical procedure. Consequently, further reduction in the time needed for the surgical procedure can be expected. With the present patients, bone grafts were made to bone defects resulting from tumor or congenital anomalies. Where the graft was made to bone that had been grafted or lengthened, the surface of the bone presented an extremely complex form, and it is likely that there would have been limited compatibility with conventional methods. Furthermore, with the present method, the CAD artificial bone data can be readily used to fashion the inner surface of the artificial bone that makes contact with the host bone, allowing this region to be shaped freely. The present method allows the surgical operator to fabricate the custom-made artificial bone into the shape that patient's desires, and this aspect of the method is likely to make it extremely useful in clinical practice. The conformity of the artificial bone during the procedure was extremely good, so that there was no need for the complicated modifications to the shape that are necessary with conventional grafts of artificial bone. The raw material used with the present method is α-TCP, and the product is a non-sintered hydroxyapatite compact. Because its resorbability is better than that of artificial bone made from a sintered body, it is expected that the present artificial bone can be replaced by bone at an earlier stage. While the present artificial bone is not as strong as artificial bone made from a sintered body, it can withstand a force of 20 MPa, and this is unlikely to be a problem for use on non-load-bearing regions of facial bone. Furthermore, since artificial bone made from a sintered body is reported to shrink by around 15% during sintering, problems remain with regard to conformity and reproducibility. With the patients in the present study, union of the artificial bone and the host bone was found in some places on the CT images starting at around 3 months after surgery. This is probably because union of the present artificial bone with host bone is more rapid than with conventional types of artificial bone, since it is highly biocompatible and non-sintered. This is a huge advantage of non-sintered artificial bone. The most important aspect of artificial bone from the point of view of the surgical procedure of reconstruction is the fixing of the artificial bone. When calcium phosphate makes contact with the surface of bone, fibrous connective tissue appears at an early stage, causing union with the bone. However, any postoperative disturbance between the artificial bone and the host bone is likely to impair union. In the present study, where a relatively large piece of artificial bone that required fixing in place was grafted, this was sutured with absorbable suture. However, a method of fixation needs to be devised in the future.

In the present cases, reconstruction was successfully performed using custom-made artificial bone, with a high level of patient satisfaction. No serious postoperative side effects attributable to the artificial bone were found, so that the safety of the artificial bone was confirmed. However, with the present fabrication method, the postoperative reaction of the patient's soft tissues depends on the intuition of the surgical operator. As even better artificial bone is developed in the future, recovery of shape to even higher degrees of accuracy and earlier bone fusion may be expected. In other words, one may anticipate the development of custom-made artificial bone designed on a system allowing prediction of the shape of the soft tissues following the artificial bone graft, and biologically active custom-made artificial bone loaded with adhesion factors. (Saijo H et al., 2009)

5. Conclusion

These artificial bones were safe and achieved dimensional compatibility along with good biodegradability and osteoconductivity. We believe that this study has provided the basis for the use of the inkjet printing technology for maxillofacial reconstruction surgery

Author details

Hideto Saijo

Oral-Maxillofacial Surgery, Dentistry and Orthodontics The University of Tokyo Hospital, Japan

References

[1] Iino M et al.,(2009). Evaluation of 15 mandibular reconstructions with Dumbach Titan Mesh-System and particulate cancellous bone and marrow harvested from bilateral posterior ilia. *Oral Surg Oral Med Oral Pathol Oral Radiol Endod* 107: 1-8, 2009

[2] Saijo H et al.,(2011). A novel method for designing and fabricating custom-made artificial bones. *Int J Oral Maxillofac Surg.* vol40, pp.955-960.

[3] Saijo H et al.,(2008). Clinical Application of Artificial Bone in the Maxillofacial Region. *J Artif Organs.* ; vol 11,pp.171–176

[4] Saijo H et al.,(2010). A bone harvesting method involving a surgical template prepared using a three-dimensional model *Asian. J. Oral.Maxillofac. Surg.* vol22,pp. 20-22

[5] Saijo H et al.,(2009). Maxillofacial reconstruction using custom-made artificial bones fabricated by inkjet printing technology. *J Artif Organs* vol.12, pp.200-205

Chapter 5

Feasibility of Molecularly Targeted Therapy for Tooth Regeneration

Katsu Takahashi, Honoka Kiso, Kazuyuki Saito,
Yumiko Togo, Hiroko Tsukamoto, Boyen Huang and
Kazuhisa Bessho

Additional information is available at the end of the chapter

http://dx.doi.org/10.5772/58904

1. Introduction

The tooth is a complex organ that consists of enamel, dentin, cementum, and pulp. Missing teeth is frequently occurring problem in aging populations. To treat these defects, the current approach involves prostheses, autotransplantation, and dental implants. The exploration of new strategies for tooth replacement has become a hot topic. Using the foundations of experimental embryology, developmental and molecular biology, tooth regeneration is becoming realistic possibility. Several different methods have been proposed to achieve biological tooth replacement. These include scaffold-based tooth regeneration, cell pellet engineering, stimulation of the formation of a third dentition, and gene-manipulated tooth regeneration. The idea that a third dentition might be locally induced to replace missing teeth is an attractive concept (Young et al., 2005; Edward & Mason, 2006; Takahashi et al., 2008, 2013). This approach is generally presented in terms of adding molecules to induce *de novo* tooth initiation in the mouth. Tooth development is the result of reciprocal and reiterative signaling between oral ectoderm-derived dental epithelium and cranial neural crest cell-derived dental mesenchyme under genetic control (Thesleff, 2006). More than 200 genes are known to be expressed during tooth development (http://bite-it.helsinki.fi/). A number of mouse mutants are now starting to provide some insights into the mechanisms of supernumerary tooth formation. Multiple supernumerary teeth may have genetic components in their etiology and partially represent the third dentition in humans. Such candidate molecules might be those that are involved in embryonic tooth induction, in successional tooth formation, or in the control of the number of teeth. This means that it may be possible to induce *de novo* tooth formation by the in situ repression or activation of a single candidate molecule. In this review, we provide an overview of the collective knowledge of tooth regeneration, especially regarding the control of the number of teeth for molecularly targeted therapy by the stimulation of a third dentition.

2. The third dentition

It has been suggested that, in humans, a "third dentition" with one or more supernumerary teeth can occur in addition to the permanent dentition, and supernumerary teeth are sometimes thought to represent a partial post-permanent dentition (Ooe, 1969). The deciduous teeth are, ontogenetically, the first generation of teeth. The permanent teeth (except molar) belong to the second dentition. The term "third dentition" refers to the opinion that one more set of teeth can occur in addition to the permanent teeth (Figure 1). Human teeth are diphyodont excepting the permanent molars. The normal mouse dentition is monophyodont and composed of one incisor and three molars in each quadrant. The number of teeth is usually strictly determined. It was initially reported that there is an anlagen of the third dentition in some mammals (Leche, 1893). The presence of an epithelial anlagen of the third dentition was also noticed in humans (Ooe, 1969). The teeth and anlagen that appear in third dentition in serial sections of infant jaws and some fetuses have been analysed. The epithelium which is considered as the anlagen of the third dentition develops lingual to all permanent tooth germs (Ooe, 1969). Furthermore, when it appears, the predecessor (permanent tooth germ) is in the bell-shaped stage (Ooe, 1969). The timing of appearance of the third dentition seems to be after birth (Table 1). This means that we have a chance to access the formation of the third dentition in the mouth.

Detailed histological analysis of the tooth replacement in these models indicates that the successional teeth are initiated from the dental lamina epithelium, which grows from the lingual side of the deciduous tooth enamel organ, and it later elongates and buds into the jaw mesenchyme, forming successional teeth. Jarvien et al. showed that, in the ferret, *Usag-1*(also known as *Sostdc1*, *Ectodin*, and *Wise*) is expressed in the elongating successional dental lamina at the interface between the lamina and deciduous tooth, as well as the buccal side of the dental lamina, suggesting that *Sostdc1* plays a role in defining the identity of the dental lamina(Jarvinen et al., 2009). Handrigan et al. analysed successional tooth formation in the snake and in lizard, and proposed that dental epithelium stem cells are responsible for the formation of successional lamina, and *Wnt* signaling may regulate the dental eithelial stem cell fate in these cells (Handrigan et al., 2010). Maintenance or reactivation of component dental lamina is thus pivotal for the replacement tooth and supernumerary formation.

3. Human syndromes associated with supernumerary teeth

Supernumerary teeth can be associated with a syndrome or they can be found in non-syndromic patients. Only 1% of non-syndromic cases have multiple supernumerary teeth, which occur most frequently in the mandibular premolar area, followed by the molar and anterior regions, respectively (Yusof, 1990; Yague-Garcia et al., 2009). Genetic mutations have been associated with the presence or absence of individual types of teeth.

Supernumerary teeth are associated with 8 syndromes and developmental abnormalities in which the resposible genes already have been isolated (Takahashi et al., 2013). The percentage occurrence in CCD is 22% in the maxillary incisor region and 5% in the molar region (Shafer, 1983). CCD is a dominantly inherited skeletal dysplasia caused by mutations in *RUNX2* (Mundlos et al., 1997). There is a wide spectrum of phenotypic variability ranging from the full-blown phenotype to an isolated dental phenotype characterized by supernumerary tooth formation and/or the delayed eruption of permanent teeth in CCD (Takahashi et al., 2008). *Runx2*-deficient mice were found to exhibit lingual buds in front of the upper molars, and these were much more prominent than in wild-type mice (Aberg et al., 2004). These buds presumably represent the mouse secondary dentition, and it is likely that *RUNX2* acts to prevent the formation of these buds. *Runx2* regulates the proliferation of cells and may exert specific control on the dental lamina and formation of successive dentitions. *Runx2* heterozygous mutant mice mostly phenocopied the skeletal defects of CCD in humans, but with no supernumerary tooth formation (Otto et al., 1997). Notably, in *Runx2* homozygous and heterozygous mouse upper molars, a prominent epithelial bud regularly presents. This epithelial bud protrudes lingually with active *Shh* signaling, and it may represent the extension of the dental lamina for successional tooth formation in mice. Hence, although *Runx2* is required for primary tooth development, it prevents the growth of the dental lamina and successional tooth formation (Otto et al., 1997).

Familial adenomatous polyposis (FAP), also named adenomatous polyposis of the colon (APC), is an autosomal dominant hereditary disorder characterized by the development of many precancerous colorectal adenomatous polyps. In addition to colorectal neoplasm, individuals can develop variable extracolonic lesions, including upper gastrointestinal polyposis, osteomas and dental anomalies (Wijn et al., 2007). Dental abnormalities include impacted teeth, congenital absence of one or more teeth, supernumerary teeth and odontomas (Wijn et al., 2007). Gardner syndrome is a variant of FAP characterized by multiple adenomas of the colon and rectum typical of FAP together with osteomas and soft tissue tumors (Chimenos-Kustner et al., 2005). Supernumerary teeth and osteomas were originally described as a part of Gardner syndrome, but they can also occur in FAP patients with or without other extracolonic lesions (Chimenos-Kustner et al., 2005; Wijn et al., 2007). FAP and Gardner syndrome are caused by a large number of germinal mutations in the *Apc* gene (Groden et al., 1991). *Apc* is a tumor suppressor gene involved in the down-regulation of free intracellular ß-catenin, the major signal transducer of the canonical Wnt signaling pathway (Groden et al., 1991). Approximately 11-27% of patients have supernumerary teeth, but, so far, no specific codon mutation of the *Apc* gene has been found to correlate with supernumerary teeth.

The identification of mutations in *RUNX2* causing an isolated dental phenotype in CCD and in *Apc* causing FAP has attracted attention as a possible route towards inducing *de novo* tooth formation.

4. Supernumerary tooth formation in a mouse model

A number of mouse mutants provide insights into the supernumerary tooth formation (Takahashi et al., 2013). Several mechanisms by which supernumerary tooth might arise in mice have been proposed (Murashima-Suginami et al., 2008; Wang et al., 2009). One plausible explanation for supernumerary tooth formation is the rescue of tooth rudiments such as within the diastema region or maxillary deciduous incisor (Yamamoto et al., 2005; Murashima-Suginami et al., 2007; Lagronova-Churava et al., 2013). During early stages of mouse tooth development transient vestigial tooth buds develop in the diastema area; developing to the bud stage yet later regressing and disappear by apoptosis, or merge with the mesial crown of the adjacent first molar tooth organ (Yamamoto et al., 2005; Lagronova-Churava et al., 2013). The rudimentary maxillary incisor regressed by apoptotic elimination of mesenchymal cells (Murashima-Suginami et al., 2007). We demonstrated that *USAG-1*-deficient mouse model has supernumerary incisors in the maxillary and mandible, a fused tooth in the maxillary and mandibular molar regions, and a supernumerary tooth was also located in front of the first mandibular molar (Figure 2). Increased *BMP* signaling results in supernumerary teeth in the *Usag-1*-deficient mouse model (Murashima-Suginami et al., 2008). Recently, we claimed that gene interactions between *BMP-7* and *USAG-1* regulate the supernumerary maxillary incisor formation (Kiso et al., 2014). *BMP-7* was co-localized with *USAG-1* in the area of the maxillary rudiment incisor tooth germ in addition to the regular maxillary incisor tooth organ. *USAG-1* abrogation rescued the apoptotic elimination of mesenchymal cells in the rudimentary maxillary incisor tooth primordia at E15, whereas the tooth sizes were comparable (Murashima-Suginami et al., 2007, 2008). The apoptotic mesenchymal cells in *USAG-1*$^{-/-}$/*BMP-7*$^{-/-}$are similar to *USAG-1*$^{+/+}$/*BMP-7*$^{+/+}$in contrast to that of *USAG-1*$^{-/-}$/*BMP-7*$^{+/+}$. These results support our interpretation that *USAG-1* functions as a novel *BMP-7* antagonist in the maxilla. We confirmed that increased *BMP* signaling in supernumerary teeth of the *USAG-1* deficient mice could be prohibited by *BMP-7* abrogation. In the contrast, to test whether *BMP-7* has the potential to induce supernumerary tooth formation, we performed explant culture and subsequent subrenal kidney capsule culture. The incisor explants supplemented with *BMP-7* in *USAG-1*$^{+/-}$as well as *USAG-1*$^{-/-}$have supernumerary tooth in similar numbers after 20 days culture, while these cultured explants in USAG-1$^{+/+}$retained normal tooth number (Figure 2). These results demonstrated that *BMP-7* can induce supernumerary tooth formation, however it is impossible to induce extra tooth by only *BMP-7* (Kiso et al., 2014). While we showed that enhanced *BMP* signaling resulted in supernumerary teeth, we also demonstarated that *BMP* signaling was modulated by *Wnt* signaling in the *Usag-1*-deficient mouse model (Figure 3) (Murashima-Suginami et al., 2008). Canonical *Wnt*/*β-catenin* signaling and its down-stream molecule Lef-1 are essential for tooth development. Overexpression of *Lef-1* under the control of the K14 promoter in transgenic mice leads to the development abnormal invaginations of the dental epithelium in the mesenchyme and formation of a tooth-like structure (Zhou et al., 1995). *De novo* supernumerary teeth arising directly from the primary tooth germ or dental lamima have been reported in *Apc* loss-of-function or *β-catenin* gain-of-function mic. It was demonstrated that mouse tooth buds expressing stabilized *β-catenin* give rise to extra teeth (Jarvinen et al., 2006). Conditional knockout of the *Apc*-gene resulted in supernumerary teeth in mice (Wang et al., 2009). Notably, adult oral tissues, especially young adult tissues, are still responsive to the loss of *Apc* (Wang et al., 2009). In old adult mice, supernumerary teeth can be induced on both labial and lingual sides of the incisors, which contain adult stem cells supporting the continuous growth of mouse incisors (Huysseune et al., 2004). In young mice, supernumerary tooth germs were induced in multiple regions of the jaw in both incisor and molar regions. They can form directly from the oral epithelium, in the dental lamina connecting the developing molar or incisor tooth germ to the oral epithelium, in the crown region, as well as in the elongating and furcation area of the developing root (Wang et al., 2009). *Wnt/BMP* signaling seems to be essential in supernumerary tooth formation (Figure 3).

We also demonstrated that *Cebpb* deficiency was related to the formation of supernumerary teeth(Figure 2). A total of 66.7% of *Cebpb*$^{-/-}$12-month-olds sustained supernumerary teeth and/or odontomas in the diastema between the incisor and the first molar (Huang et al., 2012). Furthermore, it was suggested that the dental epitheilal stem cells might be contributated to supernumerary tooth formation in mice (Figure 2).

These mouse models clearly demonstrated that it was possible to induce *de novo* tooth formation by the in situ inhibition or activation of single molecule such as *USAG-1, BMP7* or *CEBPB.*

5. Molecularly targeted therapy

Molecularly targeted therapy is a type of treatment that uses drugs or other substances to identify and attack specific types of cells by interfering with specific targeted molecules.

Most targeted therapies are either small moleclule or monoclonal antibodies. Small molecules are typically able to diffuse into cells and can act on target that are found inside the cells. Monoclonal antibodies usually can not penetorate the surface membrane and are directed against targets that are outside cells or on the surface of cells. Candidates for small molecules are identified in screing the effect of thousands of test compounds on a specific target. The best candidates are then chemically modified to produce many closely related versions. Monoclonal antibody are prepared first by immunized animals such as mice with purified target moleclules.They are humanized by replacing the animal portion of the antibody with human portion. More than 40 molecularly targeted cancer therapy have been approved by the U.S Food and Drug Administration for the treatment of specific type of cancer. Many more are in clinical trials or preclinical testing (Forscher et al., 2014). Recent molecularly targeted therapy also has successfully been introduced into the treatment of several inflammatory rheumatic diseases such as rheumatoid arthritis (Mocsai et al., 2014).

Molecularly targeted therapy provides a unique tool for the delivery of previously identified signaling molecules in both time and space that may significantly augment our progress toward clinical tooth regeneration. Stimulation of the formation of a third dentition comprises an attractive concept (Figure 4). This approach is generally presented in terms of adding molecules to induce *de novo* tooth initiation in the mouth. We have a chance to access the formation of the third dentition in the mouth, because the time of appearance of the third dentition seems to be after birth. Advances in our understanding of signal transduction by Wnt/BMP signalling in the supernumerary teeth formation offer numerous opportunities for devising new targeted therapies (Figure 3).This led to a new approach of drug development whereby targeted therapy are developed by directly targeting molecules thought to be involved in the the formation of a third dentition. A major approach for the development targeted therapeutics has been the application of monoclonal antibody for targeting molecules such as *USAG-1*.

Molecularly targeted therapy seems to be a suitable approach in tooth regeneration by stimulation of the third dentition.

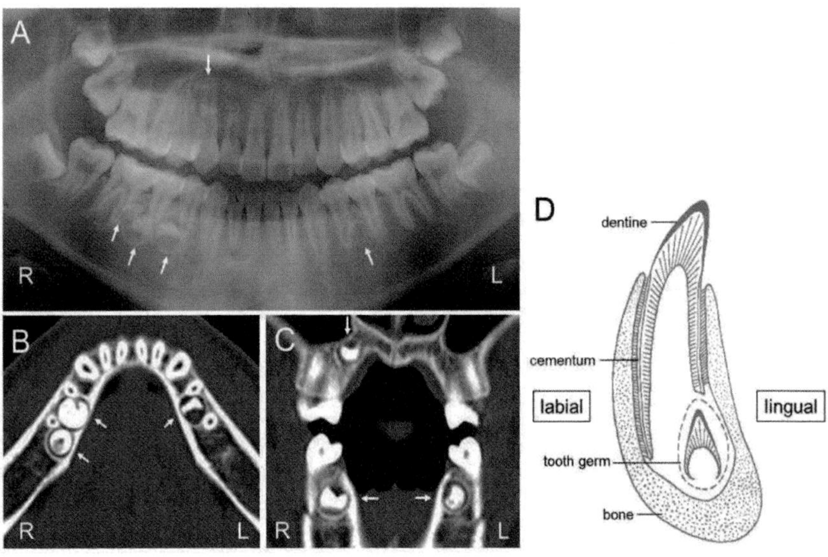

Figure 1. Multiple impacted supernumerary teeth (the third dentition) in a 13-year-old non-syndromic patient.

The third dentition develops lingual to the permanent tooth germ (D). All impacted supernumerary teeth in this patient are located to the lingual side of the permanent teeth (white arrow) (A-C). These multiple supernumerary teeth seem to be post-permanent dentition ("third dentition"). (Takahahsi et al., 2013, copy right, In Tech)

Teeth	The time of appearance of the third dentition	
	Maxilla	Mandible
central incisors	~ 3 months after birth	2 ~ 3 months after birth
lateral incisors	8 ~ 9 months after birth	2 ~ 3 months after birth
canines	2 ~ 7 months after birth	2 ~ 3 months after birth
the first premolar	1 year 1 month ~ 5 years 4 months after birth	1 year 1 month ~ 5 years 4 months after birth
the second premolar	1 year 1 month ~ 5 years 4 months after birth	2 years ~ 5 years 4 months after birth

Table 1. Timing of appearance of the third dentition (Takahahsi et al., 2013, copy right, In Tech)

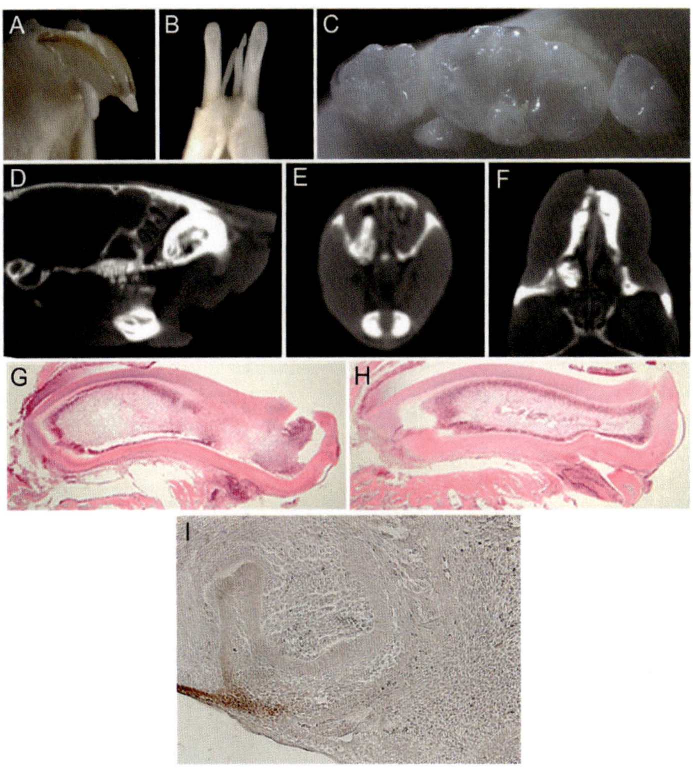

Figure 2. Supernumerary teeth formation in mouse model

Usag-1 deficient mice (A-C) A: Oblique view of the maxillary incisors. B: Occlusal view of the mandibular incisors. C: Occlusal view of the mandibular molars. Micro-CT images of *Cebpb* dificient mice (D-F) A frontal view (D), a sagittal view (E) and a horizontal view (F) *BMP-7* has potential to partially induce the formation of maxillary supernumerary incisors formation in vitro. The incisor explants supplemented with (H)/without (G) *BMP-7* in *USAG-1*$^{+/-}$ (G, H) sagittal sections of explant. (I) Sox2 positive detal epithelial stem cells in sagittal sections of the rudimentary maxillary incisor tooth primordia at E15.

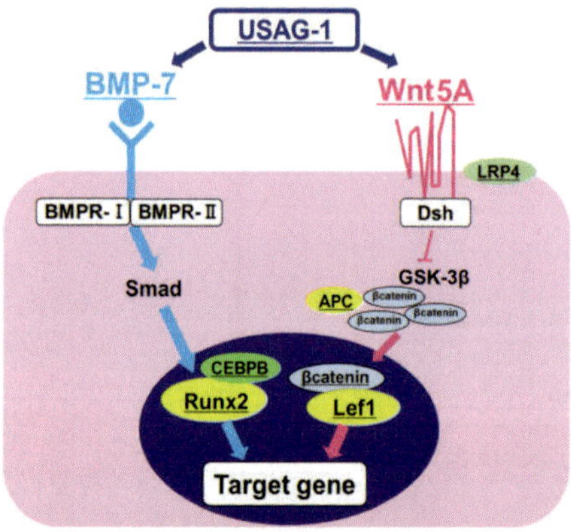

Figure 3. *Wnt/BMP* signaling in supernumerary tooth formation.

Underlined molecules are responsible genes for human syndromes or mutant mouse associated with supernumerary teeth

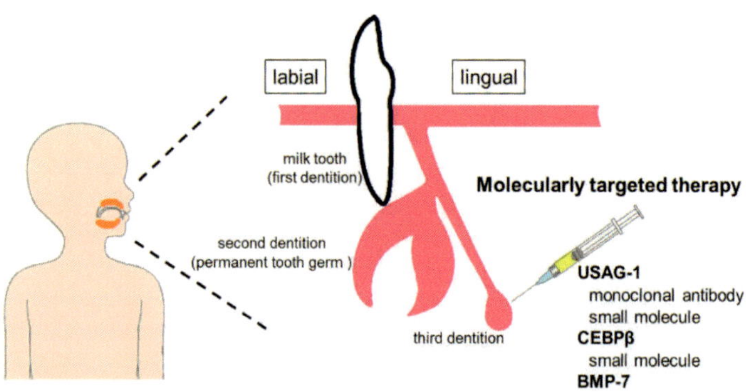

Figure 4. Molecularly targeted therapy for the tooth regeneration by stimulation of a third dentition

http://dx.doi.org/10.5772/58904

6. Conclusion

We have a chance to access the formation of the third dentition in the mouth, because the timing of the appearance of the third dentition seems to be after birth. The identification of mutations in *RUNX2* causing an isolated dental phenotype in CCD and supernumerary tooth formation in the mouse model clearly demonstrated that it was possible to induce *de novo* tooth formation by the in situ inhibition or activation of a single candidate molecule. These results support the idea that the de novo inhbition or activation of candidate molecules such as *RUNX2* or *USAG-1* might be used to stimulate the third dentition in order to induce new tooth formation in the mouse (Figure 4). Molecularly targeted therapy seems to be a suitable approach in tooth regeneration by stimulation of the third dentition.

Acknowledgements

This work was supported by Grant-in-Aid for Scientific Research(C):22592213 and 25463081 and A-STEP (Adaptable & Seamless Technology Transfer Program through Target-driven R&D) FS stage: AS231Z01061G and AS242Z02645Q.

Author details

Katsu Takahashi[1], Honoka Kiso[1], Kazuyuki Saito[1], Yumiko Togo[1], Hiroko Tsukamoto[1], Boyen Huang[2] and Kazuhisa Bessho[1]

1 Department of Oral and Maxillofacial Surgery, Graduate School of Medicine, Kyoto University, Japan

2 Department of Paediatric Dentistry, School of Medicine and Dentistry, James Cook University, Cairns, Australia

References

[1] Aberg T, Cavender A, Gaikwad JS, Bronckers AL, Wang X, Waltimo-Sirén J, Thesleff I, D'Souza RN. Phenotypic changes in dentition of Runx2 homozygote-null mutant mice. J Histochem Cytochem 2004;52(1):131-139.

[2] Chimenos-Küstner E, Pascual M, Blanco I, Finestres F. Hereditary familial polyposis and Gardner's syndrome: contribution of the odonto-stomatology examination in its diagnosis and a case description. Med Oral Patol Oral Cir Bucal 2005;10(5):402-409.

[3] Edwards PC, Mason JM. Gene-enhanced tissue engineering for dental hard tissue regeneration: (1) overview and practical considerations. Head Face Med 2006;2:12.

[4] Forscher C, Mita M, Figlin R. Targeted therapy for sarcomas. Biologics. 2014;8:91-105.

[5] Groden J, Thliveris A, Samowitz W, Carlson M, Gelbert L, Albertsen H, Joslyn G, Stevens J, Spirio L, Robertson M, et al. Identification and characterization of the familial adenomatous polyposis coli gene. Cell 1991;66(3):589-600.

[6] Handrigan GR, Leung KJ, Richman JM. Identification of putative dental epithelial stem cells in a lizard with life-long tooth replacement. Development 2010;137(21): 3545-3549.

[7] Huang B, Takahashi K, Sakata-Goto T, Kiso H, Togo Y, Saito K, Tsukamoto H, Sugai M, Akira A, Shimizu A, Bessho K. Phenotypes of CEBPB Deficiency: Supernumerary Teeth and Elongated Coronoid Process. Oral Dis 2012; 19, 144-150, 2013

[8] Huysseune A, Thesleff I. Continuous tooth replacement: the possible involvement of epithelial stem cells. Bioessays 2004;26(6):665-671.

[9] Järvinen E, Salazar-Ciudad I, Birchmeier W, Taketo MM, Jernvall J, Thesleff I. Continuous tooth generation in mouse is induced by activated epithelial Wnt/beta-catenin signaling. Proc Natl Acad Sci U S A 2006;103(49):18627-18632.

[10] Järvinen E, Tummers M, Thesleff I. The role of the dental lamina in mammalian tooth replacement. J Exp Zool B Mol Dev Evol 2009 ;312B(4):281-291.

[11] Kiso, H., Takahashi, K., Saito, K., Togo, Y., Tsukamoto, H., Huang, B., Sugai, M., Shimizu, A., Tabata, Y., Economides, AN., Slavkin, HC., Bessho, K. Interactions between BMP-7 and USAG-1 (uterine sensitization-associated gene-1) regulate supernumerary organ formations. PLoS ONE. 2014, 9, e96938, 2014

[12] Kratochwil K, Dull M, Farinas I, Galceran J, Grosschedl R. Lef1 expression is activated by BMP-4 and regulates inductive tissue interactions in tooth and hair development. Genes Dev 1996;10(11):1382-1394.

[13] Leche W. Studien uber die Entwicklung des Zahnsstemsbei den Saugetieren. Morph Jb 1893;19: 502-574.

[14] Mócsai A, Kovács L, Gergely P. What is the future of targeted therapy in rheumatology: biologics or small molecules? BMC Med. 2014;12:43.

[15] Mundlos S, Otto F, Mundlos C, Mulliken JB, Aylsworth AS, Albright S, Lindhout D, Cole WG, Henn W, Knoll JH, Owen MJ, Mertelsmann R, Zabel BU, Olsen BR. Mutations involving the transcription factor CBFA1 cause cleidocranial dysplasia. Cell 1997; 89(5): 773-779.

[16] Murashima-Suginami A, Takahashi K, Kawabata T, Sakata T, Tsukamoto H, Sugai M, Yanagita M, Shimizu A, Sakurai T, Slavkin HC, Bessho K. Rudiment incisors sur-

vive and erupt as supernumerary teeth as a result of USAG-1 abrogation. Biochem. Biophys. Res. Commun 2007;359(3):549-555.

[17] Murashima-Suginami A, Takahashi K, Sakata T, Tsukamoto H, Sugai M, Yanagita M, Shimizu A, Sakurai T, Slavkin HC, Bessho K. Enhanced BMP signaling results in supernumerary tooth formation in USAG-1 deficient mouse. Biochem Biophys Res Commun 2008;369(4):1012-1016.

[18] Ooë T. Epithelial anlagen of human third dentition and their migrations in the mandible and maxilla. Okajimas Fol Anat Jap 1969; 46(5):243-251.

[19] Otto F, Thornell AP, Crompton T, Denzel A, Gilmour KC, Rosewell IR, Stamp GW, Beddington RS, Mundlos S, Olsen BR, Selby PB, Owen MJ. Cbfa1, a candidate gene for cleidocranial dysplasia syndrome, is essential for osteoblast differentiation and bone development. Cell 1997;89(5):765-771.

[20] Shafer, W. G., Hine, M. K., and Levi, B. M. Textbook of oral pathology 4th Ed. Philadelphia:WB Saunders Co;1983.

[21] Takahashi, K., Sakata, T., Murashima-Suginami, A., Tsukamoto, H., Kiso, H. and Bessho, K. Tooth regeneration: Potential for stimulation of the formation of a third dentition by one gene. Current Topics in Genetics 2008;3: 77-82.

[22] Takahashi, K., Kiso, H., Saito, K., Togo, Y., Tsukamoto, H., Huang, B. and Bessho, K. (2013) Feasibility of gene therapy for tooth regeneration by stimulation of a third dentition: Gene Therapy-Tools and Potential Applications, In Tech, Rijeka, Croatia, 30, 727-744

[23] Thesleff I. The genetic basis of tooth development and dental defects. Am J Med Genet A 2006;140(23):2530-2535.

[24] Wang XP, O'Connell DJ, Lund JJ, Saadi I, Kuraguchi M, Turbe-Doan A, Cavallesco R, Kim H, Park PJ, Harada H, Kucherlapati R, Maas RL. Apc inhibition of Wnt signaling regulates supernumerary tooth formation during embryogenesis and throughout adulthood. Development 2009;136(11):1939-1949.

[25] Wijn MA, Keller JJ, Giardiello FM, Brand HS. Oral and maxillofacial manifestations of familial adenomatous polyposis. Oral Dis 2007;13(4):360-365.

[26] Yagüe-García J, Berini-Aytés L, Gay-Escoda C. Multiple supernumerary teeth not associated with complex syndromes: a retrospective study. Med Oral Patol Oral Cir Bucal 2009;14(7):E331-336

[27] Yamamoto H, Cho SW, Song SJ, Hwang HJ, Lee MJ, et al. (2005) Characteristic tissue interaction of the diastema region in mice. Archives of oral biology 50: 189-198.

[28] Young CS, Abukawa H, Asrican R, Ravens M, Troulis MJ, Kaban LB, Vacanti JP, Yelick PC. Tissue-engineered hybrid tooth and bone. Tissue Eng 2005;11(9-10): 1599-1610.

[29] Yusof WZ. Non-syndrome multiple supernumerary teeth: literature review. J Can Dent Assoc 1990;56(2):147-149.

[30] Zhou P, Byrne C, Jacobs J, Fuchs E. Lymphoid enhancer factor 1 directs hair follicle patterning and epithelial cell fate. Genes Dev1995;9(6):700-713.

Chapter 6

Multifaceted Neuro-Regenerative Activities of Human Dental Pulp Stem Cells for Functional Recovery after Spinal Cord Injury

Akihito Yamamoto, Kohki Matsubara, Kiyoshi Sakai,
Fumiya Kano and Minoru Ueda

Additional information is available at the end of the chapter

http://dx.doi.org/10.5772/58906

1. Introduction

Spinal cord injury (SCI) often leads to persistent functional deficits, due to loss of neurons and glia and to limited axonal regeneration after injury. Recently, three independent groups have reported that transplantation of human adult dental pulp stem cells (DPSCs) and stem cells from human exfoliated deciduous teeth (SHEDs), into the acute, sub-acute or chronic phase of rat or mouse SCI resulted in marked recovery of hindlimb locomotor functions. This review summarizes the primary characteristics of human dental pulp stem cells and their therapeutic benefits for SCI treatment. Experimental data from a number of preclinical studies suggests that pulp stem cells may promote functional recovery after SCI through multifaceted neuro-regenerative activities.

2. Dental pulp stem cells

Humans have two sets of teeth, 20 deciduous and 32 permanent ones. In the center of each tooth, there is a cavity pulp chamber, which is filled with soft connective tissue called dental pulp (Nanci and Ten Cate, 2003) (Fig.1). The major components of dental pulp are odontoblasts, fibroblasts, immune cells, extracellular matrix, blood vessels and nerve fibers. The pulp tissues are connected with systemic network through the apical foramen; this provides nutrition and sensation for responding to the external stimuli. Human adult dental pulp stem cells (DPSCs) and stem cells from human exfoliated deciduous teeth (SHEDs) are self-renewing stem cells residing within the perivascular niche of the dental pulp (Gronthos et al., 2002b). They are thought to originate from the cranial neural crest, of embryonic period and they simultaneously express early markers for both mesenchymal, neuroectodermal stem/progenitor cells and some of embryonic stem cells markers (Gronthos et al., 2000, Miura et al., 2003, Kerkis et al., 2006, Sakai et al., 2012).

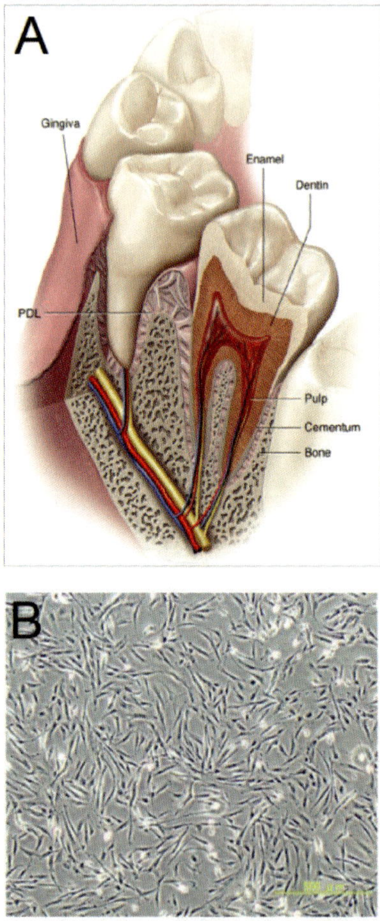

Figure 1. Diagram of tooth and pulp stem cells. (A) The tooth and its supporting structure (from Ten Cate's Oral Histlogy, Nanci and Ten Cate, 2008). PDL, Periodontal ligament. (B) Morphology of pulp stem cells. They exhibit a fibroblastic morphology with a bipolar spindle shape. Scale bar in (B): 500 μm.

Most SHEDs and DPSCs express a set of adult bone marrow stromal stem cell (BMSC) markers (CD90, CD73, and CD105), neural stem/progenitor cell markers (Doublecortin, GFAP, and Nestin), and early neuronal and oligodendrocyte markers (βIII-tubulin, A2B5 and CNPase), but not markers for mature oligodendrocytes (MBP and APC) (Sakai et al., 2012). Since naturally exfoliated deciduous and impacted adult wisdom teeth are dispensable, DPSCs and SHEDs can be easily obtained by utilizing a simple protocol (Liu et al., 2006). DPSCs and SHEDs exhibit a faster rate of proliferation and a higher number of population doublings in vitro, compared with BMSCs. Furthermore, the rate SHEDs is 1.5 times faster than that of DPSCs (Miura et al., 2003). Like BMSCs, they are multipotent cells that can differentiate in vitro into a variety of cell types including odontoblasts, osteoblasts, chondrocytes, adipocytes, endothelial cells, myocytes, and functionally active neurons (Gronthos et al., 2000, Gronthos et al., 2002a, Batouli et al., 2003, Miura et al., 2003, Nosrat et al., 2004, Kerkis et al., 2006, d'Aquino et al., 2007, Arthur et al., 2008, Arminan et al., 2009, Wang et al., 2010). Furthermore, when transplanted into the transected spinal cord (SC), they specifically differentiate toward mature oligodendrocyte lineages (Sakai et al., 2012: see below).

A cDNA microarray analysis showed that SHEDs express many genes encoding extracellular and cell-surface proteins at levels at least two-fold higher than are expressed in BMSCs (Sakai et al., 2012). It has been shown that the array of trophic factors produced by engrafted DPSCs and SHEDs provide significant therapeutic benefits for the treatment of preclinical animal disease models, including myocardial infarction, systemic lupus erythematosus (SLE), ischemic brain injury, SCI, and colitis (Gandia et al., 2008, Nakashima et al., 2009, Yamaza et al., 2010, de Almeida et al., 2011, Leong et al., 2012, Ma et al., 2012, Sakai et al., 2012, Taghipour et al., 2012, Zhao et al., 2012, Inoue et al., 2013, Yamagata et al., 2013). Thus, these studies collectively show that tooth-derived stem cells are a highly proliferative, multi-potent, and self-renewing ecto-mesenchymal stem cell-like population that actively secretes a broad repertoire of trophic and immunomodulatory factors.

3. Brief overview of the pathophysiology of SCI

The development of effective treatments for SCI has been stifled by this injury's complicated pathophysiology. During the acute phase, a primary mechanical insult disrupts tissue homeostasis. This triggers a secondary response, in which activated resident microglia and infiltrating blood-derived macrophages initiate severe inflammation by releasing high levels of multiple neurotoxic factors that induce the necrotic and apoptotic death of neurons, astrocytes, and oligodendrocytes. This response spreads beyond the initial injury site, and leads to irreversible axonal damage and demyelination (Schwab et al., 2006, Popovich and Longbrake, 2008, Rowland et al., 2008). Subsequently, reactive astrocytes and oligodendrocytes near the site of the injured spinal cord (SC) respectively produce chondroitin sulfate proteoglycans (CSPG) and myelin proteins (including myelin-associated glycoprotein (MAG), Nogo, OMG, Netrin, Semaphorin, and Ephrin). These extracellular molecules function as axon growth inhibitors (AGIs), acting through the intracellular Rho GTPase signaling cascade (Silver and Miller, 2004, Yiu and He, 2006). Thus, multiple pathogenic signals act to synergistically accelerate the progressive neuronal deterioration following SCI. Therefore, therapeutic strategies for functional recovery from SCI must exert multifaceted reparative effects targeting a variety of pathogenic mechanisms (Schwab et al., 2006).

4. Multifaceted neuro-regenerative activities of pulp stem cells

4.1. Anti-inflammatory activity

Under various pathogenic conditions, macrophages differentiate into polarized pro-inflammatory (M1) or anti-inflammatory (M2) states, and direct either detrimental or beneficial effects on tissue healing (Gordon, 2003, Mosser and Edwards, 2008). In the acute phase of SCI, the majority of accumulating microglia/macrophages are of the M1 type, and few M2 macrophages are seen throughout this period (Kigerl et al., 2009, David and Kroner, 2011). The activated M1 macrophages secrete high levels of pro-inflammatory cytokines and neurotoxic factors, including tumor necrosis factor (TNF)-α, interleukin (IL)-1β, IL-6, glutamate, and reactive oxygen species (Hausmann, 2003, Donnelly and Popovich, 2008). These neurotoxic factors accelerate glial scar formation (Popovich and Longbrake, 2008), and they induce neuronal cell death (Takeuchi et al., 2006, Block et al., 2007) and the retraction of damaged dystrophic axons (Horn et al., 2008, Busch et al., 2009). In contrast, M2 cells counteract the pro-inflammatory M1 effects and promote tissue remodeling by secreting anti-inflammatory cytokines (e.g. IL-10 and TGF-β), and scavenging cellular debris (Gordon, 2003, Mosser and Edwards, 2008, David and Kroner, 2011). Thus, macrophage polarity has the potential to determine the level of inflammation and the resultant prognosis following SCI.

Recent studies have demonstrated the induction of M2 macrophage polarization following SCI, and some of the underlying mechanisms are beginning to be elucidated. CSPG, a major component of the glial scar that is mainly known for its ability to inhibit axonal growth, has recently been shown to promote M2 polarization of infiltrating blood-derived macrophages (Rolls et al., 2008, Shechter et al., 2009, Shechter et al., 2011). In addition, recent reports have shown that BMSC transplantation using SCI or brain ischemia models leads to M2 induction (Ohtaki et al., 2008, Nakajima et al., 2012). BMSC-mediated M2 induction requires both the pre-sensitization of BMSCs by pro-inflammatory factors, such as IFN-γ, TNF-α, and LPS, and direct cell-to-cell contact (Nemeth et al., 2009, Singer and Caplan, 2011). Thus, CSPG together with pro-inflammatory factors in the injured SC may be involved in the pre-sensitization of engrafted BMSCs to activate their M2-inducing machinery.

As described in the previous section, SHEDs also exhibit strong immunosuppressive properties that effectively ameliorate several autoimmune diseases, including SLE and colitis (Yamaza et al., 2010, Ma et al., 2012, Zhao et al., 2012). Importantly, intravenously administered SHEDs express Fas-Ligand, which induces T-cell apoptosis, thereby triggering immune tolerance (Zhao et al., 2012). This elevates the ratio of regulatory T cells (Tregs) to pro-inflammatory T cells, resulting in anti-inflammatory conditions (Yamaza et al., 2010). We also found that, in the mouse hypoxic ischemia model, both intracerebral transplantation of SHEDs, and administration of serum-free conditioned media (CM) derived from SHEDs (SHED-CM), generates anti-inflammatory conditions and promotes functional recovery (Yamagata et al., 2013). Thus, tooth-derived stem cells have strong immunoregulatory properties that promote tissue regeneration in the injured CNS.

4.2. Regeneration of the injured axon

Both axonal regeneration and the re-formation of appropriate neuronal connections are required for functional recovery from SCI. However, multiple AGIs block the inherent regenerative capacities of injured axons (Silver and Miller, 2004, Schwab et al., 2006, Yiu and He, 2006, Rowland et al., 2008). It is well known that AGIs constitute an intricate molecular network in the extracellular space of the injured CNS, where they activate a common intracellular signaling mediator, Rho GTPase, and its effector, Rho-associated kinase (ROCK) (Maekawa et al., 1999, Winton et al., 2002, Dubreuil et al., 2003, Monnier et al., 2003, Yamashita and Tohyama, 2003). Activation of the Rho-ROCK cascade induces growth-cone collapse and axonal repulsion (Hall, 1998). In contrast, inactivation of either Rho by C3 transferase, or ROCK by the kinase inhibitor Y-27632 down-regulates AGI signaling and promotes functional recovery after SCI (Lehmann et al., 1999, Dergham et al., 2002, Fournier et al., 2003). Thus, Rho-ROCK signaling is an important target for SCI treatments; however, few studies have investigated the effect of stem-cell transplantation on regulating AGI/Rho-ROCK signaling cascades.

Importantly, engrafted SHEDs were recently shown to promote the regeneration of two major types of descending axons (CST and 5-HT) beyond the lesion epicenter, and to concomitantly inhibit SCI-induced Rho activation. Furthermore, both SHED-CM and DPSC-CM (but not BMSC-CM) promote neurite extension by primary cerebral granular neurons (CGNs) cultured on two different AGIs (CSPG and MAG) (Sakai et al., 2012). Thus, tooth-derived stem cells promote the regeneration of transected axons through the direct inhibition of multiple AGI signals by paracrine mechanisms.

In addition, the engraftment of DPSCs into avian embryos results in the chemoattraction of trigeminal ganglion axons via the chemokine CXCL12 and its receptor, CXCR4 (Arthur et al., 2009). DPSCs and SHEDs express several neurotropic factors that promote neurite extension (de Almeida et al., 2011, Sakai et al., 2012). Our preliminary analysis showed that these trophic factors, when applied individually, failed to promote the neurite extension of CGNs cultured on CSPG-coated dishes; however it is possible that they may promote axonal regeneration in a synergistic manner.

4.3. Anti-apoptotic activity

Pharmacological blockade of neuron and/or oligodendrocyte apoptosis by a number of agents promotes functional recovery after SCI. These agents include the following: erythropoietin (Celik et al., 2002, Gorio et al., 2002), inhibitors of purine receptor P2X7 (OxATP and PPADS) (Wang et al., 2004), a neutralizing antibody against CD95 (FAS) antigen (Demjen et al., 2004), and minocycline (Stirling et al., 2004, Teng et al., 2004). Engrafted SHEDs suppress the apoptosis of neurons and oligodendrocytes, resulting in the remarkable preservation of neurofilaments and myelin sheaths in the region surrounding the lesion epicenter (Nosrat et al., 2001, de Almeida et al., 2011). Intracerebral transplantation of DPSCs from rhesus macaques promotes proliferation, cell recruitment, and maturation of endogenous stem/progenitor cells by modulating the local microenvironment (Huang et al., 2008). Notably, SHEDs also strongly inhibit the apoptosis of astrocytes recruited to the lesion (Sakai et al., 2012).

Classically, reactive, CSPG-generating astrocytes have been considered an obstacle to axonal regeneration; however, recent genetic studies in mice indicate that the conditional ablation of astrocytes after SCI results in larger lesions, failure of blood-brain-barrier repair, increased inflammation and tissue disruption, severe demyelination, and profound cell death of neurons and oligodendrocytes (Bush et al., 1999, Faulkner et al., 2004, Okada et al., 2006, Herrmann et al., 2008, Rolls et al., 2009). Thus, the collective evidence demonstrates that, in addition to their anti-regenerative activity, astrocytes also play an important role in neuro-protection during the acute phase of SCI. SHEDs can suppress astrocyte apoptosis and minimize secondary injury, as well as inhibit the AGI activity of CSPG derived from astrocytes. Thus, SHEDs have the potential to promote functional recovery after SCI through two distinct mechanisms involving astrocyte regulation.

4.4. Cell-replacement activity

Undifferentiated rat and human pulp stem cells can form neurospheres in vitro (Sasaki et al., 2008, Wang et al., 2010) and simultaneously express multiple neural stem/progenitor markers (Gronthos et al., 2002a, Miura et al., 2003, Sakai et al., 2012). In addition, DPSCs can differentiate in vitro toward functionally active neurons, which express voltage-gated Na+channels, and in vivo toward neuron-like cells 48 hours after transplantation into the mesencephalon of avian embryos (Arthur et al., 2008). Furthermore, simultaneous PKC and cAMP activation induces the differentiation of DPSCs into functionally active neurons (Kiraly et al., 2009). Thus, pulp stem cells display a capacity for neuronal differentiation both in vivo and in vitro.

Recently, three independent groups reported that pulp stem cells show neuro-regenerative activity in rodent SCI models. Interestingly, engrafted pulp stem cells promoted significant functional recovery in all three studies, but exhibited variable capacities for differentiation. In the first study, DPSCs were transplanted into the compressed mouse SC at day 7 (sub-acute phase) or day 28 (chronic phase) after injury, and the engrafted DPSCs differentiated into glia cells expressing S-100 and GFAP (de Almeida et al., 2011). In the second study, undifferentiated or neural-phenotype induced SHED (iSHED) were transplanted into the contused rat SC at 7 days after injury. Engrafted SHED and iSHED differentiated primarily into MAP2+mature neurons and GFAP+astrocytes, and to a lesser extent into MBP-and NG2-expressing oligo-dendrocytes (Taghipour et al., 2012). In the third study, from our group, undifferentiated SHEDs were transplanted into the completely transected rat SC immediately after the surgery. The engrafted SHEDs survived well following SCI: more than 30% of the engrafted SHEDs survived as a cell mass in the injured SC 8 weeks after transplantation and more than 90% of the engrafted SHEDs differentiated toward mature oligodendrocytes, expressing APC and MBP (Sakai et al., 2012).

http://dx.doi.org/10.5772/58906

Taken together, these experimental data suggest that the microenvironment of the transplanted stem cells significantly affects their capacity for differentiation. In the acute phase of SCI, the injured SC contains high levels of pro-inflammatory mediators. Thus these factors may activate the oligodendrocyte-specific differentiation cascade of pulp stem cells.

5. Conclusion

Recent experimental data from a number of studies reveals that engrafted SHEDs provide a number of distinct therapeutic benefits for treatment of SCI: (1) the suppression of the early inflammatory response; (2) inhibition of the SCI-induced apoptosis of neurons, astrocytes, and oligodendrocytes, which promotes the preservation of neural fibers and myelin sheaths; (3) regeneration of the transected axon through the direct inhibition of multiple AGI signals (including CSPG and MAG) by paracrine mechanisms; and (4) cell replacement in the damaged SC through the SHEDs' capacity for differentiation towards oligodendrocytes, neurons and astrocytes. Thus, we propose that tooth-derived stem cells may provide significant therapeutic benefits for treating SCI through both cell-autonomous and paracrine/trophic regenerative activities.

Acknowledgements

This work was supported by grants from Ministry of Education, Culture, Sports, Science and Technology.

Author details

Akihito Yamamoto, Kohki Matsubara, Kiyoshi Sakai, Fumiya Kano and Minoru Ueda

Nagoya University Graduate School of Medicine, Japan

References

[1] Arminan A, Gandia C, Bartual M, Garcia-Verdugo JM, Lledo E, Mirabet V, Llop M, Barea J, Montero JA, Sepulveda P (2009) Cardiac differentiation is driven by NKX2.5 and GATA4 nuclear translocation in tissue-specific mesenchymal stem cells. Stem Cells Dev 18:907-918.

[2] Arthur A, Rychkov G, Shi S, Koblar SA, Gronthos S (2008) Adult Human Dental Pulp Stem Cells Differentiate Toward Functionally Active Neurons Under Appropriate Environmental Cues. Stem Cells 26:1787-1795.

[3] Arthur A, Shi S, Zannettino ACW, Fujii N, Gronthos S, Koblar SA (2009) Implanted adult human dental pulp stem cells induce endogenous axon guidance. Stem Cells 27:2229-2237.

[4] Batouli S, Miura M, Brahim J, Tsutsui TW, Fisher LW, Gronthos S, Robey PG, Shi S (2003) Comparison of stem-cell-mediated osteogenesis and dentinogenesis. J Dent Res 82:976-981.

[5] Block ML, Zecca L, Hong JS (2007) Microglia-mediated neurotoxicity: uncovering the molecular mechanisms. Nat Rev Neurosci 8:57-69.

[6] Busch S, Horn K, Silver D, Silver J (2009) Overcoming Macrophage-Mediated Axonal Dieback Following CNS Injury. J Neurosci 29:9967.

[7] Bush TG, Puvanachandra N, Horner CH, Polito A, Ostenfeld T, Svendsen CN, Mucke L, Johnson MH, Sofroniew MV (1999) Leukocyte infiltration, neuronal degeneration, and neurite outgrowth after ablation of scar-forming, reactive astrocytes in adult transgenic mice. Neuron 23:297-308.

[8] Celik M, Gokmen N, Erbayraktar S, Akhisaroglu M, Konakc S, Ulukus C, Genc S, Genc K, Sagiroglu E, Cerami A, Brines M (2002) Erythropoietin prevents motor neuron apoptosis and neurologic disability in experimental spinal cord ischemic injury. Proc Natl Acad Sci USA 99:2258-2263.

[9] d'Aquino R, Graziano A, Sampaolesi M, Laino G, Pirozzi G, De Rosa A, Papaccio G (2007) Human postnatal dental pulp cells co-differentiate into osteoblasts and endotheliocytes: a pivotal synergy leading to adult bone tissue formation. Cell Death Differ 14:1162-1171.

[10] David S, Kroner A (2011) Repertoire of microglial and macrophage responses after spinal cord injury. Nat Rev Neurosci 12:388-399.

[11] de Almeida FM, Marques SA, Ramalho Bdos S, Rodrigues RF, Cadilhe DV, Furtado D, Kerkis I, Pereira LV, Rehen SK, Martinez AM (2011) Human dental pulp cells: a new source of cell therapy in a mouse model of compressive spinal cord injury. J Neurotrauma 28:1939-1949.

[12] Demjen D, Klussmann S, Kleber S, Zuliani C, Stieltjes B, Metzger C, Hirt U, Walczak H, Falk W, Essig M, Edler L, Krammer P, Martin-Villalba A (2004) Neutralization of CD95 ligand promotes regeneration and functional recovery after spinal cord injury. Nat Med 10:389-395.

[13] Dergham P, Ellezam B, Essagian C, Avedissian H, Lubell WD, McKerracher L (2002) Rho signaling pathway targeted to promote spinal cord repair. J Neurosci 22:6570-6577.

[14] Donnelly DJ, Popovich PG (2008) Inflammation and its role in neuroprotection, axonal regeneration and functional recovery after spinal cord injury. Exp Neurol 209:378-388.

[15] Dubreuil C, Winton M, McKerracher L (2003) Rho activation patterns after spinal cord injury and the role of activated Rho in apoptosis in the central nervous system. J Cell Biol 162:233-243.

[16] Faulkner JR, Herrmann JE, Woo MJ, Tansey KE, Doan NB, Sofroniew MV (2004) Reactive astrocytes protect tissue and preserve function after spinal cord injury. J Neurosci 24:2143-2155.

[17] Fournier AE, Takizawa BT, Strittmatter SM (2003) Rho kinase inhibition enhances axonal regeneration in the injured CNS. J Neurosci 23:1416-1423.

[18] Gandia C, Arminan A, Garcia-Verdugo JM, Lledo E, Ruiz A, Minana MD, Sanchez-Torrijos J, Paya R, Mirabet V, Carbonell-Uberos F, Llop M, Montero JA, Sepulveda P (2008) Human dental pulp stem cells improve left ventricular function, induce angiogenesis, and reduce infarct size in rats with acute myocardial infarction. Stem Cells 26:638-645.

[19] Gordon S (2003) Alternative activation of macrophages. Nature Reviews Immunology 3:23-35.

[20] Gorio A, Gokmen N, Erbayraktar S, Yilmaz O, Madaschi L, Cichetti C, Di Giulio A, Vardar E, Cerami A, Brines M (2002) Recombinant human erythropoietin counteracts secondary injury and markedly enhances neurological recovery from experimental spinal cord trauma. Proc Natl Acad Sci USA 99:9450-9455.

[21] Gronthos S, Brahim J, Li W, Fisher LW, Cherman N, Boyde A, DenBesten P, Robey PG, Shi S (2002a) Stem cell properties of human dental pulp stem cells. J Dent Res 81:531-535.

[22] Gronthos S, Brahim J, Li W, Fisher LW, Cherman N, Boyde A, Denbesten P, Robey PG, Shi S (2002b) Stem Cell Properties of Human Dental Pulp Stem Cells. J Dent Res 81:531-535.

[23] Gronthos S, Mankani M, Brahim J, Robey PG, Shi S (2000) Postnatal human dental pulp stem cells (DPSCs) in vitro and in vivo. Proc Natl Acad Sci USA 97:13625-13630.

[24] Hall A (1998) Rho GTPases and the actin cytoskeleton. Science 279:509-514.

[25] Hausmann ON (2003) Post-traumatic inflammation following spinal cord injury. Spinal Cord 41:369-378.

[26] Herrmann JE, Imura T, Song B, Qi J, Ao Y, Nguyen TK, Korsak RA, Takeda K, Akira S, Sofroniew MV (2008) STAT3 is a critical regulator of astrogliosis and scar formation after spinal cord injury. J Neurosci 28:7231-7243.

[27] Horn K, Busch S, Hawthorne A, Van Rooijen N, Silver J (2008) Another Barrier to Regeneration in the CNS: Activated Macrophages Induce Extensive Retraction of Dystrophic Axons through Direct Physical Interactions. J Neurosci 28:9330.

[28] Huang AH-C, Snyder BR, Cheng P-H, Chan AWS (2008) Putative dental pulp-derived stem/stromal cells promote proliferation and differentiation of endogenous neural cells in the hippocampus of mice. Stem Cells 26:2654-2663.

[29] Inoue T, Sugiyama M, Hattori H, Wakita H, Wakabayashi T, Ueda M (2013) Stem cells from human exfoliated deciduous tooth-derived conditioned medium enhance recovery of focal cerebral ischemia in rats. Tissue engineering Part A 19:24-29.

[30] Kerkis I, Kerkis A, Dozortsev D, Stukart-Parsons GC, Gomes Massironi SM, Pereira LV, Caplan AI, Cerruti HF (2006) Isolation and characterization of a population of immature dental pulp stem cells expressing OCT-4 and other embryonic stem cell markers. Cells Tissues Organs 184:105-116.

[31] Kigerl KA, Gensel JC, Ankeny DP, Alexander JK, Donnelly DJ, Popovich PG (2009) Identification of two distinct macrophage subsets with divergent effects causing either neurotoxicity or regeneration in the injured mouse spinal cord. The Journal of neuroscience 29:13435-13444.

[32] Kiraly M, Porcsalmy B, Pataki A, Kadar K, Jelitai M, Molnar B, Hermann P, Gera I, Grimm WD, Ganss B, Zsembery A, Varga G (2009) Simultaneous PKC and cAMP activation induces differentiation of human dental pulp stem cells into functionally active neurons. Neurochem Int 55:323-332.

[33] Lehmann M, Fournier A, Selles-Navarro I, Dergham P, Sebok A, Leclerc N, Tigyi G, McKerracher L (1999) Inactivation of Rho signaling pathway promotes CNS axon regeneration. J Neurosci 19:7537-7547.

[34] Leong WK, Henshall TL, Arthur A, Kremer KL, Lewis MD, Helps SC, Field J, Hamilton-Bruce MA, Warming S, Manavis J, Vink R, Gronthos S, Koblar SA (2012) Human adult dental pulp stem cells enhance poststroke functional recovery through non-neural replacement mechanisms. Stem cells translational medicine 1:177-187.

[35] Liu H, Gronthos S, Shi S (2006) Dental pulp stem cells. Methods Enzymol 419:99-113.

[36] Ma L, Makino Y, Yamaza H, Akiyama K, Hoshino Y, Song G, Kukita T, Nonaka K, Shi S, Yamaza T (2012) Cryopreserved dental pulp tissues of exfoliated deciduous teeth is a feasible stem cell resource for regenerative medicine. PLoS One 7:e51777.

[37] Maekawa M, Ishizaki T, Boku S, Watanabe N, Fujita A, Iwamatsu A, Obinata T, Ohashi K, Mizuno K, Narumiya S (1999) Signaling from Rho to the actin cytoskeleton through protein kinases ROCK and LIM-kinase. Science 285:895-898.

[38] Miura M, Gronthos S, Zhao M, Lu B, Fisher LW, Robey PG, Shi S (2003) SHED: stem cells from human exfoliated deciduous teeth. Proc Natl Acad Sci USA 100:5807-5812.

[39] Monnier PP, Sierra A, Schwab JM, Henke-Fahle S, Mueller BK (2003) The Rho/ROCK pathway mediates neurite growth-inhibitory activity associated with the chondroitin sulfate proteoglycans of the CNS glial scar. Mol Cell Neurosci 22:319-330.

[40] Mosser DM, Edwards JP (2008) Exploring the full spectrum of macrophage activation. Nat Rev Immunol 8:958-969.

[41] Nakajima H, Uchida K, Rodriguez Guerrero A, Watanabe S, Sugita D, Takeura N, Yoshida A, Long G, Wright K, Johnson E, Baba H (2012) Transplantation of Mesenchymal Stem Cells Promotes the Alternative Pathway of Macrophage Activation and Functional Recovery after Spinal Cord Injury. J Neurotrauma.

[42] Nakashima M, Iohara K, Sugiyama M (2009) Human dental pulp stem cells with highly angiogenic and neurogenic potential for possible use in pulp regeneration. Cytokine Growth Factor Rev 20:435-440.

[43] Nanci A, Ten Cate AR (2003) Ten Cate's oral histology : development, structure, and function. St. Louis, Mo. ; London: Mosby.

[44] Nanci A, Ten Cate AR (2008) Ten Cate's oral histology : development, structure, and function. St. Louis, Mo.: Mosby.

[45] Nemeth K, Leelahavanichkul A, Yuen PS, Mayer B, Parmelee A, Doi K, Robey PG, Leelahavanichkul K, Koller BH, Brown JM, Hu X, Jelinek I, Star RA, Mezey E (2009) Bone marrow stromal cells attenuate sepsis via prostaglandin E(2)-dependent reprogramming of host macrophages to increase their interleukin-10 production. Nat Med 15:42-49.

[46] Nosrat IV, Smith CA, Mullally P, Olson L, Nosrat CA (2004) Dental pulp cells provide neurotrophic support for dopaminergic neurons and differentiate into neurons in vitro; implications for tissue engineering and repair in the nervous system. Eur J Neurosci 19:2388-2398.

[47] Nosrat IV, Widenfalk J, Olson L, Nosrat CA (2001) Dental pulp cells produce neurotrophic factors, interact with trigeminal neurons in vitro, and rescue motoneurons after spinal cord injury. Dev Biol 238:120-132.

[48] Ohtaki H, Ylostalo JH, Foraker JE, Robinson AP, Reger RL, Shioda S, Prockop DJ (2008) Stem/progenitor cells from bone marrow decrease neuronal death in global ischemia by modulation of inflammatory/immune responses. Proc Natl Acad Sci U S A 105:14638-14643.

[49] Okada S, Nakamura M, Katoh H, Miyao T, Shimazaki T, Ishii K, Yamane J, Yoshimura A, Iwamoto Y, Toyama Y, Okano H (2006) Conditional ablation of Stat3 or Socs3 discloses a dual role for reactive astrocytes after spinal cord injury. Nat Med 12:829-834.

[50] Popovich PG, Longbrake EE (2008) Can the immune system be harnessed to repair the CNS? Nat Rev Neurosci 9:481-493.

[51] Rolls A, Shechter R, London A, Segev Y, Jacob-Hirsch J, Amariglio N, Rechavi G, Schwartz M (2008) Two faces of chondroitin sulfate proteoglycan in spinal cord repair: a role in microglia/macrophage activation. PLoS Med 5:e171.

[52] Rolls A, Shechter R, Schwartz M (2009) The bright side of the glial scar in CNS repair. Nat Rev Neurosci 10:235-241.

[53] Rowland JW, Hawryluk GWJ, Kwon B, Fehlings MG (2008) Current status of acute spinal cord injury pathophysiology and emerging therapies: promise on the horizon. Neurosurg Focus 25:E2.

[54] Sakai K, Yamamoto A, Matsubara K, Nakamura S, Naruse M, Yamagata M, Sakamoto K, Tauchi R, Wakao N, Imagama S, Hibi H, Kadomatsu K, Ishiguro N, Ueda M (2012) Human dental pulp-derived stem cells promote locomotor recovery after complete transection of the rat spinal cord by multiple neuro-regenerative mechanisms. The Journal of clinical investigation 122:80-90.

[55] Sasaki R, Aoki S, Yamato M, Uchiyama H, Wada K, Okano T, Ogiuchi H (2008) Neurosphere generation from dental pulp of adult rat incisor. Eur J Neurosci 27:538-548.

[56] Schwab JM, Brechtel K, Mueller CA, Failli V, Kaps HP, Tuli SK, Schluesener HJ (2006) Experimental strategies to promote spinal cord regeneration--an integrative perspective. Prog Neurobiol 78:91-116.

[57] Shechter R, London A, Varol C, Raposo C, Cusimano M, Yovel G, Rolls A, Mack M, Pluchino S, Martino G, Jung S, Schwartz M (2009) Infiltrating blood-derived macrophages are vital cells playing an anti-inflammatory role in recovery from spinal cord injury in mice. PLoS Med 6:e1000113.

[58] Shechter R, Raposo C, London A, Sagi I, Schwartz M (2011) The glial scar-monocyte interplay: a pivotal resolution phase in spinal cord repair. PLoS One 6:e27969.

[59] Silver J, Miller JH (2004) Regeneration beyond the glial scar. Nat Rev Neurosci 5:146-156.

[60] Singer NG, Caplan AI (2011) Mesenchymal stem cells: mechanisms of inflammation. Annu Rev Pathol 6:457-478.

[61] Stirling DP, Khodarahmi K, Liu J, McPhail LT, McBride CB, Steeves JD, Ramer MS, Tetzlaff W (2004) Minocycline treatment reduces delayed oligodendrocyte death, attenuates axonal dieback, and improves functional outcome after spinal cord injury. J Neurosci 24:2182-2190.

[62] Taghipour Z, Karbalaie K, Kiani A, Niapour A, Bahramian H, Nasr-Esfahani MH, Baharvand H (2012) Transplantation of undifferentiated and induced human exfoliated deciduous teeth-derived stem cells promote functional recovery of rat spinal cord contusion injury model. Stem Cells Dev 21:1794-1802.

[63] Takeuchi H, Jin S, Wang J, Zhang G, Kawanokuchi J, Kuno R, Sonobe Y, Mizuno T, Suzumura A (2006) Tumor necrosis factor-alpha induces neurotoxicity via glutamate

release from hemichannels of activated microglia in an autocrine manner. J Biol Chem 281:21362-21368.

[64] Teng YD, Choi H, Onario RC, Zhu S, Desilets FC, Lan S, Woodard EJ, Snyder EY, Eichler ME, Friedlander RM (2004) Minocycline inhibits contusion-triggered mitochondrial cytochrome c release and mitigates functional deficits after spinal cord injury. Proc Natl Acad Sci USA 101:3071-3076.

[65] Wang J, Wang X, Sun Z, Wang X, Yang H, Shi S, Wang S (2010) Stem cells from human-exfoliated deciduous teeth can differentiate into dopaminergic neuron-like cells. Stem Cells Dev 19:1375-1383.

[66] Wang X, Arcuino G, Takano T, Lin J, Peng W, Wan P, Li P, Xu Q, Liu Q, Goldman S, Nedergaard M (2004) P2X7 receptor inhibition improves recovery after spinal cord injury. Nat Med 10:821-827.

[67] Winton MJ, Dubreuil CI, Lasko D, Leclerc N, McKerracher L (2002) Characterization of new cell permeable C3-like proteins that inactivate Rho and stimulate neurite outgrowth on inhibitory substrates. J Biol Chem 277:32820-32829.

[68] Yamagata M, Yamamoto A, Kako E, Kaneko N, Matsubara K, Sakai K, Sawamoto K, Ueda M (2013) Human dental pulp-derived stem cells protect against hypoxic-ischemic brain injury in neonatal mice. Stroke 44:551-554.

[69] Yamashita T, Tohyama M (2003) The p75 receptor acts as a displacement factor that releases Rho from Rho-GDI. Nat Neurosci 6:461-467.

[70] Yamaza T, Kentaro A, Chen C, Liu Y, Shi Y, Gronthos S, Wang S, Shi S (2010) Immunomodulatory properties of stem cells from human exfoliated deciduous teeth. Stem cell research & therapy 1:5.

[71] Yiu G, He Z (2006) Glial inhibition of CNS axon regeneration. Nat Rev Neurosci 7:617-627.

[72] Zhao Y, Wang L, Jin Y, Shi S (2012) Fas ligand regulates the immunomodulatory properties of dental pulp stem cells. J Dent Res 91:948-954.

Chapter 7

Potential Cell-Based Therapies for Irreversibly Damaged Salivary Glands and Atrophic Alveolar Bone

Yoshinori Sumita, Izumi Asahina, Simon D. Tran,
Hideki Agata, Minoru Inoue, Arinobu Tojo and
Hideaki Kagami

Additional information is available at the end of the chapter

http://dx.doi.org/10.5772/58913

1. Introduction

One of the major recent advances in medicine is the use of living cells for treatment. For example, hematopoietic stem cells (HSCs) have been transplanted to leukemic patients (bone marrow transplantation) (McGovern et al., 1956; Atkinson et al., 1959; Thomas et al., 1959). Isolated stem cells from the donor were infused into the recipient's blood stream. Those infused cells replaced the recipient's diseased HSCs. Similarly, pancreatic islets were isolated from donors and transplanted to diabetic patients (Shapiro et al., 2000). In these applications, the cells were isolated and used directly without modifications. From the 1980's, this approach has been expanded to include the use of cells that have been extensively manipulated *in vitro*. For example, autologous keratinocytes were isolated from skin, expanded *in vitro* and transplanted to the skin defects of burned patients (Gallico et al., 1984). A similar approach has been reported for the replacement of damaged cartilage at the knee joint. Cartilage cells harvested from non-load bearing site of the knee joint were cultured and transplanted to the damaged site, leading to a significant reduction of pain and increased mobility (Brittberg et al., 1994). Currently, treatments using living cells (either non-modified or modified cells) are one of the most promising fields in medicine. As for the field of dentistry, treatments using living cells have been investigated for several tissues. In particular, bone defects and damaged salivary glands have been considered as realistic targets for cell-based therapies (Kagami et al., 1998, 2014; Tran et al., 2011). This chapter will focus on the history and current status of cell therapy to those dental organs.

2. Cell therapy to salivary gland

Patients with the irreversible loss of salivary gland function are seen in Sjögren's syndrome (SS) and after radiation ± chemotherapy treatment for head and neck cancer. They usually suffer considerable morbidity and severe reduction in their quality of life because the dysfunction of the salivary glands leads to severe xerostomia (dry mouth), dysphagia, dental caries, oro-pharyngeal infections and diminished mucosal wound healing. Although the etiopathologic bases of SS and radiogenic atrophy are quite different, these conditions arise from the progressive loss of acinar cells, which are the principal site of fluid-and protein-secretion in salivary glands. Unfortunately, there are no adequate treatments for patients with such irreversible glandular damage. Current pharmacological approaches aim to increase the secretory capacity of the surviving acinar cells but this approach is not feasible if few or no acinar cells remain in the glands. Therefore, developing alternative treatment strategies to restore acinar cells in damaged salivary glands are required.

2.1. Experimental approaches to restore the functional salivary glands

Recently, experimental approaches to regenerate functional acinar cells such as the use of gene therapy, tissue engineering (for developing an artificial gland), or cell-based therapy have been explored with the aim of developing novel clinical treatments (Baum & Tran, 2006; Baum et al., 2012; Khalili et al., 2014; Sugito et al., 2004; Sumita et al., 2011; Tran et al., 2005). Gene therapy involves the transfer of genes into residual cells of atrophied glands to promote saliva secretion, and it is a promising strategy for a future clinical treatment. This approach focuses on the delivery of a water-channel protein gene to the surviving ductal epithelial cells using a recombinant adenovirus vector. In fact, promising results were recently reported from a clinical trial administering the *aquaporin 1* gene for radiogenic dysfunction of salivary glands (Baum et al., 2012). An alternate strategy, tissue engineering an artificial salivary gland, is also a promising approach to the replacement of lost or damaged glands. This strategy requires both salivary epithelial stem/progenitor cells and a biodegradable scaffold that reconstructs the microenvironment of glandular tissues. We have shown that it is feasible to culture salivary epithelial cells for their eventual use in a prototype artificial salivary gland (Tran et al., 2005). However, though this strategy can generate one portion of salivary parenchymal tissue (ductal cells), it is difficult to regenerate the fully functional salivary tissues (both ductal and acinar cells). Recently, a fully functional salivary gland in adult mice has been shown possible through the orthotopic transplantation of a bioengineered salivary gland germ, which was reconstituted from epithelial and mesenchymal cells isolated from an embryonic salivary gland germ (Ogawa et al., 2013). This study provided valuable results for future tissue engineering strategies to create an artificial salivary gland in patients.

Meanwhile using another experimental strategy, we have focused on a cell-based therapy with the intention to deliver stem/progenitor cells to atrophic glands by intra-glandular or –venous injections for increasing the regenerative capacity of the damaged tissues and providing the salivary gland stem cells that can differentiate to functional acinar cells. To date, promising cell sources using this strategy have been stem/progenitor cells derived from salivary glands, bone marrow, umbilical cord, or adipose tissue (Khalili et al., 2012, 2014; Kojima et al., 2011; Lombaert et al., 2008; Sugito et al, 2004; Sumita et al, 2011; Xu et al, 2012). Using cells from the salivary glands, we previously demonstrated that the cultured rat salivary epithelial cells could be transplanted into atrophic glands. These cells remained in the damaged gland tissues for 4 weeks (Sugito et al., 2004). Moreover, it was shown that the spheroid culture of adult salivary gland cells could enrich the progenitor cells (salispheres) and these progenitors restored the gland morphology and functions after intra-glandular transplantation (Lombaert et al., 2008). These reports showed that the stem/progenitor cells derived from salivary glands could be considered as a promising source of cell therapy for the radiogenic dysfunction of salivary glands. Our group has also focused on bone marrow-derived cells (BMDCs), including mesenchymal stem cells (MSCs) as a cell-based strategy. We previously found that donor BMDCs migrated to salivary glands and transdifferentiated into salivary epithelial cells after their intravenous injection in human patients (Tran et al., 2011). Owing to this, their potential benefits would have been expected to regenerate the functions of atrophic glands caused by both SS and radiation therapy. It has been reported that BMDCs display the effects of paracrine, vasculogenesis, transdifferentiation or immunomodulation after transplantation in the regeneration of various tissues (Tran et al., 2011). Herein, as a prerequisite for future clinical trials, we summarize our preliminary results of studies using two different models of damage to salivary glands, which allowed to analyze the regenerative capacity of BMDCs.

2.2. Behaviors of donor BMDCs in radiogenic dysfunction model

Treatment for most patients with head and neck cancers includes ionizing radiation ± chemotherapy. And, this treatment causes the irreversible damage to salivary glands, which is accompanied with a loss of fluid-secreting acinar cells and a considerable decrease of saliva secretion. To develop a cell-based therapy for this dysfunction, we firstly investigated whether non-cultured fresh BMDCs could differentiate into salivary epithelial cells and restore gland's function in head and neck irradiated mice (Sumita et al., 2011). BMDCs from male donor mice were transplanted through the tail-vein of female recipient mice post gamma-ray irradiation of 15 or 18 Gy. After 8, 16, and 24 weeks, saliva secretion was increased in mice treated by BMDCs transplantation (Figure 1).

At 24 weeks after irradiation, harvested submandibular-and parotid-glands of BMDC-treated mice had greater weights than those of non-treated mice, and possessed an increased level of tissue regenerative activity (blood vessel formation, cell proliferation, and epidermal growth factor activity), while apoptosis activity was increased in non-treated salivary glands. The expression of stem cell markers (Sca-1 or c-Kit) was increased in BMDC-treated salivary glands. Additionally, we found an increased area of acinar cells and approximately 9% of Y-chromosome positive salivary epithelial cells (derived from the donor) in BMDC-treated mice. Therefore, we drew a conclusion from this study that cell therapy using BMDCs could rescue the functional damage of irradiated salivary glands through mechanisms involving a paracrine effect as well as a direct differentiation of BMDCs into salivary epithelial cells. Likewise, several studies have shown the beneficial effects of cell-based therapies using MSCs derived from bone marrow or adipose tissues, to date (Kojima et al., 2011; Lim et al., 2013). In addition, as a proof-

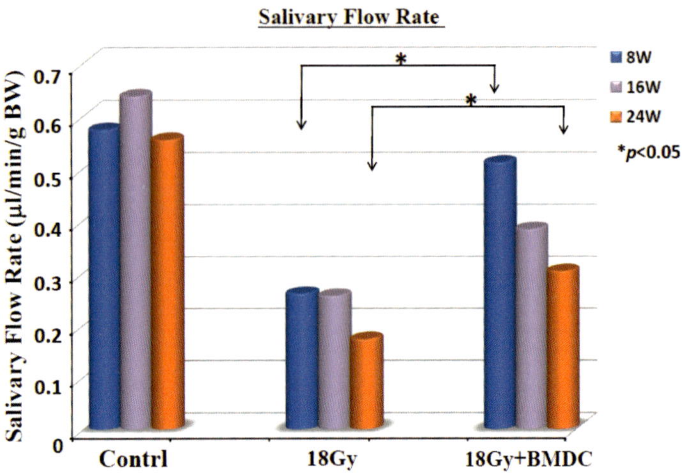

(Modified from Sumita et al., 2011)

Figure 1. Salivary flow rate recovery after BMDCs transplantation at 8, 16, 24 weeks post-irradiation. SFR were higher in BMDC-transplanted mice (18Gy+BMDC) when compared to non-transplanted ones (18Gy).

of-concept of the paracrine effect of (BMDCs) cell-based therapy, our group recently demonstrated that the injection of bone marrow cell-extracts (a bioactive lysate of BMDCs) could be advantageously used to repair the radiogenic dysfunction of salivary glands rather than BMDCs transplantation (Tran et al., 2013). Although the mechanisms of regeneration are not well understood at the present, these findings provided a promising result for future clinical trials of cell-based therapy using BMDCs or MSCs.

2.3. Behaviors of donor BMDCs in SS model

Sjögren's syndrome (SS) is a chronic autoimmune disease characterized by infiltrates of lymphocytes in the salivary glands. In this disease, the immune system attacks the salivary glands, particularly the acinar cells. This causes an irreversible damage that is accompanied with a progressive loss of fluid-secreting acinar cells, and leads to a considerable decrease of saliva secretion. The challenge of successful cell-based therapy for autoimmune diseases is not only to regenerate the tissue but also to prevent it from the same autoimmune attack that was responsible for its destruction at the first place (Kodama et al., 2003). Following the above-mentioned concept, we initially copied a therapy proposed by Faustman and colleagues to reverse end-stage diabetes in Non-obese diabetic (NOD) mouse, but for the treatment of SS-like disease (Tran et al., 2007). NOD mice develop SS-like disease and a progressive loss of saliva secretion. This therapy had two components. First, it was an injection of complete Freund's adjuvant (CFA) to induce endogenous TNF-α to exterminate autoreactive T lymphocytes. The second component was the transplantation of MHC class I-matched normal spleen cells. As a result, our group showed that injections of CFA combined with spleen cells restored salivary function in NOD mice with SS-like disease (Tran et al., 2007). However for future clinical consideration, spleen cells are not easily obtained from patients. Therefore our group modified the Faustman's two-component therapy by using BMDCs (Khalili et al., 2010). The spleen and bone marrow are closely related organs, and both are among the first sites of hematopoiesis during gestation. We injected CFA and MHC class I-matched normal BMDCs in 7-week-old NOD mice, which had not yet developed SS. At 52 weeks post-treatment, we found that all NOD mice receiving BMDCs and CFA had a recovery of saliva secretion and were protected from SS and diabetes. Although a very small number of donor BMDCs had differentiated into salivary epithelial cells (<0.1%), including acinar cells was observed, the size of focus score (the number of foci, infiltrated lymphocytes) was bigger in the non-treated mice (Figure 2).

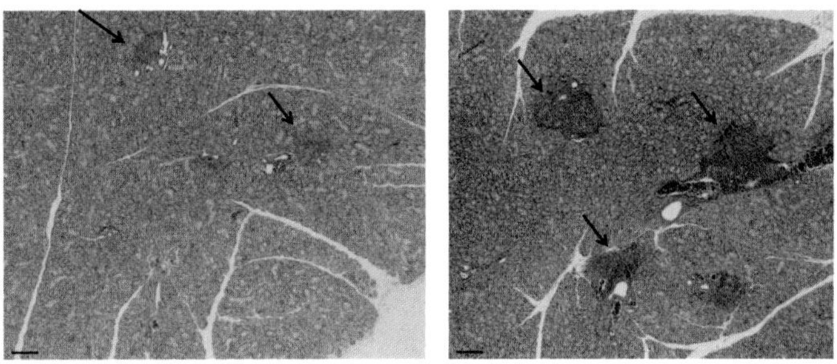

A: BMDC-treated NOD NOD

B: Non-treated

(Modified from Khalili et al., 2010)

Figure 2. Focus score change after BMDCs transplantation. The size of the focus score (black arrows) was bigger in non-treated mice (Non-treated NOD) when compared to BMDC-treated mice (BMDC-treated NOD).

This study suggests that a combined immuno and cell-based therapy can permanently prevent SS and restore the salivary function in NOD mice through mechanisms involving the paracrine and immunomodulatory effects. Moreover, we analyzed whether this BMDC-treatment would be effective in restoring salivary gland function if the treatment was given at a late phase of SS (20-week old NOD) when minimal saliva was secreted. In consequence, saliva secretion improved and was at best 50 % of pre-symptomatic levels. The treatment also decreased TNF-α and TGF-β1 levels while increasing EGF and regulatory T cells. This study provided that the cell-based therapy using BMDCs holds a promising effect, even when given in an advanced phase of SS. Likewise, we have also confirmed MSCs derived from bone marrow display similar effectiveness (Khalili et al., 2012). The combined use of CFA and MSCs was effective in both preventing saliva secretion loss and reducing lymphocytic influx in salivary glands. Thus, cell-based therapy must be recognized as one of the promising options for future clinical trials. Recently, our group has also shown that the injection of bone marrow cell extracts (a bioactive lysate of BMDCs) could restore the function of damaged salivary gland in NOD mice (Misuno et al., 2014).

3. Cell-based therapy to atrophic alveolar bone (bone tissue engineering)

Atrophic alveolar bone is one of the major obstacles for dental implant therapy because dental implant placement requires the presence of a minimal amount of bone. For patients with severe bone atrophy, functional recovery with conventional dentures is difficult and most of the patients suffer from masticatory disturbance. A combination of autologous bone graft and dental implant installation is a realistic option for those patients with severe alveolar bone atrophy. However, the procedure is accompanied by swelling and pain of the donor site and causes morbidity. Bioartificial bone substitutes have been frequently used as an alternative, although the artificial materials cannot induce bone regeneration and the application is limited. Accordingly, tissue engineering and regenerative medicine of bone tissue is now receiving significant attention (Kagami et al., 2014).

3.1. Scientific bases for bone tissue engineering

Osteogenic ability is the function to generate bone after transplantation. Before the development of tissue-engineered bone, only vascular bone grafts were known to have this function. Tissue-engineered bone is designed to contain living osteogenic cells and thus possesses osteogenic function. The characteristic features of tissue-engineered bone is based on the presence of osteogenic cells.

As a source for osteogenic cells, pluripotent mesenchymal stromal cells from bone marrow aspirates have been widely used. Initially, this group of cells was defined as adherent fibroblast-like cells in bone marrow. Since the culture contains cells which can generate single cell-derived colonies, those cells were referred to as colony forming units-fibroblasts (CFUs-F). Eventually, CFUs-F were proved to have high proliferating potential and even single-colony derived cells were shown capable of forming bone (Friedenstein et al., 1966; Friedenstein et al. 1970; Friedenstein et al., 1987). Subsequently, the capability of CFUs-F to differentiate to various mesenchymal tissues has been reported. Then, these cells were re-named as"mesenchymal stem cells" (MSCs) (Caplan, 1991). However, not all cells are multipotent; MSCs are also designated as "bone marrow stromal cells (BMSCs)", which is a relatively widely accepted nomenclature (Prockop, 2009). Although BMSCs can be obtained from a patient's own bone marrow aspirate, the percentage of BMSCs in total bone marrow cells is likely less than 0.01% (Pittenger et al., 1999). Accordingly, cell cultivation is required to obtain a sufficient number of cells for clinical bone tissue engineering. For example, bone marrow from a single rat can provide a sufficient number of osteogenic cells for only two or three bone-forming transplants. In contrast, after *in vitro* expansion of cells from the same volume of bone marrow, more than 50 bone-forming transplants are possible (Yoshikawa et al., 1996).

The quality of cells is an important factor for tissue engineering as it could affect the treatment outcome. Accordingly, BMSCs should be appropriately processed, cultured and induced into

http://dx.doi.org/10.5772/58913

osteogenic cells. For this purpose, various parameters including cell separation procedures, culture media, type and origin of serum, cell seeding density, timing of passaging, the number of passages, content and concentrations of reagents for induction and the period of induction should be optimized. Similarly, scaffold material and shape should be optimized depending on the purpose. To establish a standard operating protocol for the preparation of cells for clinical bone tissue engineering, we have investigated optimal cell culture and induction procedures using human BMSCs and granular type β-tricalcium phosphate (β-TCP) as a scaffold. The results showed that the passage number, seeding density and the period of induction significantly affected the osteogenic ability of BMSCs (Agata et al., 2010). In particular, human BMSCs lose their *in vivo* bone forming ability very rapidly after passaging and no bone formation was observed with cells after the fourth cell passage (Fig. 3) (Agata et al., 2010). The results of those basic studies should be considered in the establishment of an optimal cell culture/induction protocol.

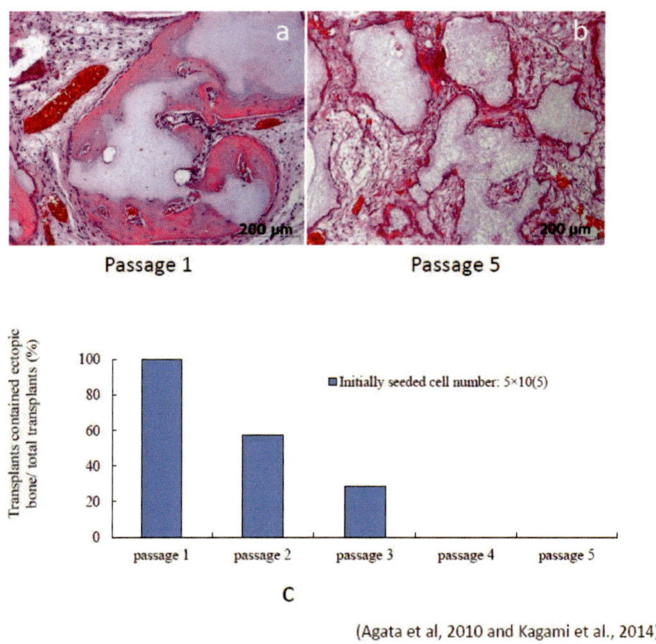

Figure 3. Effect of passage numbers on BMSC's ability for ectopic bone formation. Upper panels show ectopic bone formation on the back of nude mice with tissue-engineered bone using human BMSCs at passage 1 (a) and passage 5 (b). The success of ectopic bone formation quickly decrease after cell passage and no bone formation was observed after passage 4 (c). Note that the bone forming ability was quickly lost during passage. Adapted from Kagami et al., 2014 and modified from original figure in Agata et al., 2010.

In terms of the safety of bone tissue engineering, BMSCs have been used clinically for more than ten years and no major side effect has been reported (Quatro et al., 2001). Accordingly, bone tissue engineering using BMSCs is considered relatively safe. However, genetic instability has been reported in mouse BMSCs, changes that might lead to carcinogenesis (reviewed by Prockop, 2009). Quest for the usage of more potent stem cells might be accompanied by a higher risk of genetic alteration, an issue that should be kept in mind in therapeutic applications of stem cells.

3.2. Clinical bone tissue engineering in dentistry and maxillofacial surgery

Tissue-engineered bone was first applied for long-bone defects (Quatro et al., 2001). In Quatro's study, autologous BMSCs were cultured with hydroxyapatite blocks and transplanted to the defective site. This study was the first to prove the clinical feasibility of bone tissue engineering.

In the field of dentistry, results from a first clinical study were reported in 2004. MSCs were mixed with platelet rich plasma as a scaffold and transplanted to the site of atrophic alveolar bone at the time of dental implant installation (Yamada et al., 2004). They reported that bone regeneration was observed in all cases. More recently, another clinical study using BMSCs was reported (Meijer et al., 2008). BMSCs were cultured and seeded on hydroxyapatite granules. Then, the cells were induced into osteogenic cells for one week and transplanted. In this study, the subjects included severe atrophy cases in which dental implant could not be installed at the time of cell transplantation (two steps approach). Bone formation was observed in three cases. However, in two cases where the atrophy was severe, newly formed bone was only observed in the area adjacent to the native bone, which might imply osteoconduction rather than bone regeneration by the transplanted cells. Accordingly, the efficacy of clinical alveolar bone tissue engineering for severe atrophy cases was not established in this study.

We have conducted a clinical study of bone tissue engineering for severe atrophy of alveolar bone. BMSCs were harvested from each patient's bone marrow, expanded *in vitro*, induced into osteogenic cells and transplanted together with β-TCP granules (Fig. 4). The results following two-and five-year observations proved that bone regeneration was successful and no side effects or related complications have been observed. On the other hand, one of the important findings through this clinical study was the presence of individual variations in cell growth, differentiation and levels of bone regeneration (Asahina et al., manuscript in preparation). This problem is not limited to bone tissue engineering and might be an important issue for the all therapies using autologous cells. Development of novel technologies to reduce individual variation is a current research target.

3.3. Future prospect of bone tissue engineering in dentistry and maxillofacial surgery

Currently, various techniques for bone regeneration/reconstruction are available, including microvascular-free flaps, autologous block bone transplantation, allogeneic or xenogeneic bone substitutes or purely artificial bone substitutes. Furthermore, recent development of additional strategies such as destruction osteogenesis, growth factor administration and tissue

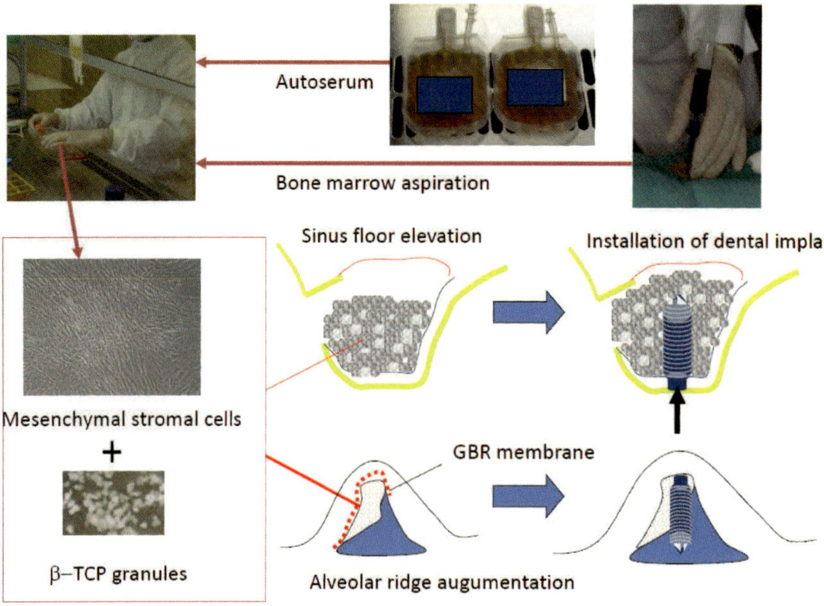

Figure 4. The procedure for clinical study of alveolar bone regeneration at The Research Hospital, The Institute of Medical Science, The University of Tokyo. Modified from original figure in Kagami et al., 2014.

engineering will provide more options. Accordingly, the selection of an adequate technique for each case is of importance for practitioners.

When the tissue defect is relatively large and the condition of the recipient site is not optimal, the use of more invasive technique such as autologous bone grafting might be acceptable. However, future bone regeneration therapy will probably avoid surgical removal of healthy bone in favour of less invasive procedures. Bone tissue engineering is one of the most promising approaches and is expected to replace cases of autologous bone grafting. Currently, the best applications and most significant limitations of bone tissue engineering are not clear. Well-designed clinical studies should be performed to answer these important questions, and these clinical trials will contribute to the wide acceptance of bone tissue engineering.

4. Conclusion

In the near future, a cell-based therapy will be accepted as a useful or even an essential treatment option for various diseases in dentistry and maxillofacial surgery. Based on previous and ongoing studies, radiation-damaged salivary glands and atrophic alveolar bone are the most realistic targets. Despite rapid progress in studies of cell-based therapies, this novel technology has yet to gain acceptance in ordinary practice. Since cell-based therapy is still expensive, determination of adequate treatment targets should be an important research goal to facilitate its widespread use. Furthermore, it is essential to reduce treatment costs without sacrificing safety. Development of automated cell culture systems and efficient safety tests are the key technologies to achieve this goal.

Acknowledgements

This work was supported in part by "Health Labour Sciences Research Grants" from the Ministry of Health, Labor and Welfare of Japan,"Grant-in-Aid for Scientific Research (B)", "Grant-in-Aid for challenging Exploratory Research" from Japan Society for the Promotion of Science (JSPS) and "Translational Research Network Program" from Ministry of Education, Culture, Sports, Science and Technology (MEXT) of Japan.

Author details

Yoshinori Sumita[1], Izumi Asahina[1], Simon D. Tran[2], Hideki Agata[1], Minoru Inoue[3,4], Arinobu Tojo[3] and Hideaki Kagami[3,4]

1 Department of Regenerative Oral Surgery, Unit of Translational Medicine, Graduate School of Biomedical Sciences, Nagasaki University, Nagasaki, Japan

2 Laboratory of Craniofacial Tissue Engineering, Faculty of Dentistry, McGill University, Montreal, Quebec, Canada

3 Tissue Engineering Research Group, Division of Molecular Therapy, The Advanced Clinical Research Center, The Institute of Medical Science, The University of Tokyo, Tokyo, Japan

4 Department of Oral and Maxillofacial Surgery, Matsumoto Dental University, Shiojiri, Japan

References

[1] Agata, H., Asahina, I., Watanabe, N., Ishii, Y., Kubo, N., Ohshima, S., Yamazaki, M., Tojo, A. &Kagami, H. (2010). *Characteristic change and loss of in vivo osteogenic abilities of human bone marrow stromal cells during passage.* Tissue Eng Part A, Vol.16, No.2,(October 2010), pp. 663–673, ISSN 1937-3341 (Print).

[2] Atkinson, J.B., Mahoney, F.J., Schwartz, I.R. & Hesch, J.A. (1959). *Therapy of acute leukemia by whole-body irradiation and bone marrow transplantation from an identical normal twin.* Blood. Vol. 14, No. 3, (March 1959), pp. 228-234, ISSN 0006-4971 (Print).

[3] Baum BJ, Tran SD. Synergy between genetic and tissue engineering – creating an artificial salivary gland. *Periodontology2000.* 2006;41:218-223

[4] Baum, B.J., Alevizos, I., Zheng, C., Cotrim, A.P., Liu, S., McCullagh, L., Goldsmith, C.M., Burbelo, P.D., Citrin, D.E., Mitchell, J.B., Nottingham, L.K., Rudy, S.F., Van Waes, C., Whatley, M.A., Brahim, J.S., Chiorini, J.A., Danielides, S., Turner, R.J., Patronas, N.J., Chen, C.C., Nikolov, N.P. & Illei, G.G. (2012). *Early responses to adenoviral-mediated transfer of the aquaporin-1 cDNA for radiation-induced salivary hypofunction.* Proc Natl Acad Sci USA, Vol. 109, No. 47, (November 2012), pp. 19403-19407, ISSN: 0027-8424 (Print).

[5] Brittberg, M., Lindahl, A., Nilsson, A., Ohlsson, C., Isaksson, O. & Peterson, L. (1994), *Treatment of deep cartilage defects in the knee with autologous chondrocyte transplantation.* N Engl J Med, Vol. 331, No. 14, (October 1994), pp. 889-895, ISSN 0021-9738 (Print).

[6] Caplan, A.I. (1991), *Mesenchymal stem cells.* J Orthop Res, (September 1991), Vol. 9, No. 5, pp. 641–650, ISSN:0736-0266 (Print).

[7] Friedenstein, A.J., Piatetzky-Chapiro, I.I. & Petrakova, K.V. (1966), *Osteogenesis in transplants of bone marrow cells.* J Embryol Exp Morph, (December 1966), Vol.16, No. 3, pp. 381–390, ISSN:0022-0752 (Print).

[8] Friedenstein, A.J., Chailakhjan, R.K. & Lalykina, K.S. (1970). *The development of fibroblast colonies in monolayer cultures of guinea-pig bone marrow and spleen cells.* Cell Tissue Kinet (October 1970), Vol. 3, No. 4, pp.393–403, ISSN:0008-8730 (Print).

[9] Friedenstein, A.J., Challakhyan, R.K., Gerasimov, U.V. (1987). *Bone marrow osteogenic stem cells: in vitro cultivation and transplantation in diffusion chambers.* Cell Tissue Kinet, (May 1987), Vol. 20, No. 3, pp. 263–272, ISSN:0008-8730 (Print).

[10] Gallico, G. G. 3rd, O'Connor, N. E., Compton, C. C., Kehinde, O. & Green, H. (1984). Permanent Coverage of Large Burn Wounds with Autologous Cultured Human Epithelium. *N Engl J Med,* Vol. 311, No. 7, (August 1984), pp. 448-451, ISSN 0028-4793.

[11] Kagami, H., Wang, S. & Hai, B. (2008). *Restoring the function of salivary glands.* Oral Dis. Vol. 14, No. 1, (January 2008), pp. 15-24, ISSN 1354-523X (Print).

[12] Kagami, H., Agata, H., Inoue, M., Asahina, I., Tojo, A., Yamashita, N. & Imai, K. (2014). *The Use of Bone Marrow Stromal Cells (Bone Marrow-Derived Multipotent Mesenchymal Stromal Cells) for Alveolar Bone Tissue Engineering: Basic Science to Clinical Translation.* In press, (2014), ISSN 1937-3368 (Print).

[13] Khalili, S., Liu, Y., Sumita, Y., Maria, O.M., Blank, D., Key, S., Mezey, E. & Tran, S.D. (2010). *Bone marrow cells are a source of undifferentiated cells to prevent Sjogren's syn-*

drome and to preserve salivary glands function in the non-obese diabetic mice. Int J Biochem Cell Biol, (November 2010), Vol. 42, No. 11, pp. 1893-1899, ISSN:1357-2725 (Print).

[14] Khalili, S., Liu, Y., Kornete, M., Roescher, N., Kodama, S., Peterson, A., Piccirillo, C.A. & Tran S.D. (2012). *Mesenchymal stromal cells improve salivary function and reduce lymphocytic infiltrates in mice with Sjogren's-like disease.* PLoS One, (June, 2012), Vol. 7, No. 6, pp.e38615, ISSN:1932-6203 (Electronic).

[15] Khalili, S., Faustman DL, Liu, Y., Sumita, Y., Blank, D., Peterson, A., Kodama, S. & Tran, S.D. (2014). *Treatment for salivary gland hypofunction at both initial and advanced stages of Sjogren-like disease: a comparative study of bone marrow therapy versus spleen cell therapy with a 1-year monitoring period.* Cytotherapy, (March 2014), Vol. 6, No. 3, pp. 412-423, ISSN:1465-3249 (Print).

[16] Kodama, S., Kuhtreiber, W., Fujimura, S., Dale, E.A. & Faustman, D.L. (2003). *Islet regeneration during the reversal of autoimmune diabetes in NOD mice.* Science, (November 2003), Vol. 302, No. 5648, pp. 1223-1227, ISSN:0036-8075 (Print).

[17] Kojima, T., Knemaru, S.-I., Hirano, S., Tateya, I., Ohno, S., Nakamura, T. & Ito, J. (2011). *Regeneration of radiation damaged salivary glands with adipose-derived stromal cells.* Laryngoscope, (September 2011), Vol. 121, No. 9, pp. 1864-1869, ISSN:0023-852X (Print).

[18] Lim, J.Y., Yi, T., Choi, J.S., Jang, Y.H., Lee, S., Kim, H.J., Song, S.U. & Kim, Y.M. (2013). *Intraglandular transplantation of bone marrow-derived clonal mesenchymal stem cells for amelioration of post-irradiation salivary gland damage.* Oral Oncol, (February 2013), Vol. 49, No. 2, pp. 136-143, ISSN:1368-8375 (Print).

[19] Lombaert, I.M., Brunsting, J.F., Wierenga, P.K., Faber,, H., Stokman, M.A., Kok, T., Visser, W.H., Kampinga, H.H., de Haan, G. & Coppes R.P. (2008). *Rescue of salivary gland function after stem cell transplantation in irradiated glands.* PLoS One, (April 2008), Vol. 3, No. 4, pp. e2063, ISSN:1932-6203 (Electronic).

[20] Meijer, G.J., de Bruijn, J.D., Koole, R. & van Blitterswijk, C.A. (2008). *Cell based bone tissue engineering in jaw defects.* Biomaterials, (July 2008), Vol. 29, No. 21, pp. 3053-3061, ISSN:0142-9612 (Print).

[21] McGovern, J.J. Jr., Russell, P.S., Atkins, L., Webster, E.W. (1956). *Treatment of terminal leukemic relapse by total-body irradiation and intravenous infusion of stored autologous bone marrow obtained during remission.* N Engl J Med. Vol. 260, No. 16, (April 1956), pp. 675-683, ISSN 0028-4793 (Print).

[22] Misuno, K., Tran, S.D., Khalili, S., Huang, J., Liu, Y. & Hu, S. (2014). *Quantitative analysis of protein and gene expression in salivary glands of Sjorgren's-like disease NOD mice treated by bone marrow soup.* PLoS One, (January 2014), Vol. 9, No. 1, pp. e87158, ISSN: 1932-6203 (Electronic).

[23] Ogawa, M., Oshima, M., Imamura, A., Sekine, Y., Ishida, K., Yamashita, K., Nakajima, K., Hirayama, M., Tachikawa, T. & Tsuji, T. (2013). *Functional salivary gland regen-*

eration by transplantation of a bioengineered organ germ. Nat Commun, Vol. 4, pp. 2498, ISSN:2041-1723 (Electronic).

[24] Pittenger, M.F., Mackay, A.M., Beck, S.C., Jaiswal, R.K., Douglas, R., Mosca, J.D., Moorman, M.A., Simonetti, D.W., Craig, S. & Marshak, D.R. (1999). *Multilineage potential of adult human mesenchymal stem cells.* Science (April 1999), Vol. 284, No. 5411, pp. 143–147, ISSN:0036-8075 (Print).

[25] Prockop DJ. (2009). *Repair of tissues by adult stem/progenitor cells (MSCs): controversies, myths, and changing paradigms.* Mol Ther, (June 2009), Vol. 17, No. 6, pp. 939–946, ISSN:1525-0016 (Print).

[26] Quatro, R., Mastrogiacomo, M. & Cancedda, R. (2001). *Repair of large bone defects with the use of autologous bone marrow stromal cells.* N Engl J Med (February 2001), Vol. 344, No. 5, pp. 385–386, ISSN:0028-4793 (Print).

[27] Shapiro, A.M., Lakey, J.R., Ryan, E.A., Korbutt, G.S., Toth, E., Warnock, G.L., Kneteman, N.M. & Rajotte, R.V. (2000). *Islet transplantation in seven patients with type 1 diabetes mellitus using a glucocorticoid-free immunosuppressive regimen.* N Engl J Med. Vol. 343, No. 4, (July 2000), pp. 230-238, ISSN 0028-4793.

[28] Sugito, T., Kagami, H., Hata, K., Nishiguchi, H. & Ueda, M. (2004). *Transplantation of cultured salivary gland cells into an atrophic salivary gland.* Cell Transplant, Vol. 13, No. 6, pp. 691-699, ISSN:0963-6897 (Print).

[29] Sumita, Y., Liu, Y., Khalili, S., Maria, O.M., Xia, D., Key, S., Cotrim, A.P., Mezey, E. & Tran, S.D. (2011). *Bone marrow-derived cells rescue salivary gland function in mice with head and neck irradiation.* Int J Biochem Cell Biol, (January 2011), Vol. 43, No. 1, pp. 80-87, ISSN:1357-2725 (Print).

[30] Thomas, E.D., Lochte, H.L. Jr., Cannon, J.H., Sahler, O.D., Ferrebee, J.W. (1959). *Supralethal whole body irradiation and isologous marrow transplantation in man.* J Clin Invest. (October 1959), Vol. 38, No. 10 (Pt 1-2), pp. 1709-1716, ISSN 0021-9738 (Print).

[31] Tran, S.D., Wang, J., Bandyopadhyay, B.C., Redman, R.S., Dutra, A., Pak, E., Swaim, W.D., Gerstenhaber, J.A., Bryant, J.M., Zheng, C., Goldsmith, C.M., Kok, M.R., Wellner, R.B. & Baum, B.J. (2005). *Primary culture of polarized human salivary epithelial cells for use in developing an artificial salivary gland.* Tissue Eng, (January-February 2005), Vol. 11, No. 1-2, pp. 172-181, ISSN:1076-3279 (Print).

[32] Tran, S.D., Kodama, S., Lodde, B.M., Szalayova, I., Key, S., Khalili, S., Faustman, D.L. & Mezey, E. (2007). *Reversal of Sjogren's-like syndrome in non-obese diabetic mice.* Ann Rheum Dis, (June 2007), Vol. 66, No. 6, pp. 812-814, ISSN:0003-4967 (Print).

[33] Tran, S.D., Sumita, Y. & Khalili, S. (2011). *Bone marrow-derived cells: A potential approach for the treatment of xerostomia.* Int J Biochem Cell Biol. (October 2010), Vol. 43, No. 1, pp. 5-9, ISSN 1357-2725 (Print).

[34] Tran, S.D., Redman, R.S., Barrett, A.J., Pavletic, S.Z., Key, S., Liu, Y., Carpenter, A., Nguyen, H.M., Sumita, Y., Baum, B.J., Pillemer, S.R. & Mezey, E. (2011). *Microchimer-*

ism in salivary glands after blood- and marrow-derived stem cell transplantation. Biol Blood Marrow Transplant, (March, 2011), Vol. 17, No. 3, pp. 429-433, ISSN:1083-8791 (Print).

[35] Tran, S.D., Liu, Y., Xia, D., Maria, O.M., Khalili, S., Wang, R.W., Quan, V.H., Hu, S. & Seuntjens, J. (2013). *Paracrine effects of bone marrow soup restore organ function, regeneration, and repair in salivary glands damaged by irradiation.* PLoS One, (April 2013), Vol. 8, No 4, pp. e61632, ISSN:1932-6203 (Electronic).

[36] Yamada, Y., Ueda, M., Hibi, H. & Nagasaka, T. (2004). *Translational research for injectable tissue-engineered bone regeneration using mesenchymal stem cells and platelet-rich plasma: from basic research to clinical case study.* Cell Transplant, Vol. 13, No. 4, pp. 343–355, ISSN:0963-6897 (Print).

[37] Yoshikawa, T., Ohgushi, H. & Tamai, S. (1996). *Immediate bone forming capability of prefabricated osteogenic hydroxyapatite.* J Biomed Mater Res (November 1996), Vol. 32, No. 3, pp. 481–492, ISSN:0021-9304 (Print).

[38] Xu, J., Xu, J., Wang, D., Liu, D., Fan, Z., Zhang, H., Liu, O., Ding, G., Gao, R., Zhang, C., Ding, Y., Bromberg, J.S., Chen, W., Sun, L. & Wang, S. (2012). *Allogeneic mesenchymal stem cell treatment alleviates experimental and clinical Sjogren's syndrome.* Blood. (October 2011), Vol. 120, No. 15, pp. 3142-3151, ISSN:0006-4971 (Print).

Chapter 8

Bone and Periodontal Tissue Regeneration by the Secretomes from Bone Marrow Derived Mesenchymal Stem Cells

Wataru Katagiri

Additional information is available at the end of the chapter

1. Introduction

Alveolar bone regeneration with grafting is often carried out prior to placement of dental implants. Several graft materials have been used including autogenous bone, xenogeneic bone, and synthetic bone substitutes. Autogenous bone grafts have been used for a long time with good predictability and are considered the "gold standard" because of their osteoinductive and osteoconductive properties and immunogenic compatibility. However, autogenous bone must be harvested from a donor site of the patient and is associated with higher morbidity (Arrington et al., 1996; Joshi et al., 2004). Xenogeneic bone and synthetic bone substitutes such as deproteinized bovine bone, hydroxyapatite, and calcium triphosphate are often used clinically as osteoconductive scaffolds, but they provide limited osteoinductivity and a potential risk of infection and extrusion (Damien et al., 1991). Osteoinductive growth factors such as bone morphogenic protein (BMP)-2 have been used with these osteoconductive materials to promote bone regeneration (Herford & Boyne, 2008). However, recent studies have indicated unexpected effects on bone regeneration including induction of a severe inflammatory response, because of the higher dose with clinical application of BMP-2 (Kawasaki et al., 1998; Perri et al., 2007, Vaidya et al., 2007).

Recently, the concept of tissue engineering and regenerative medicine has been widely accepted (Langer & Vacanti, 1993), and many clinical studies have been performed including studies of bone and periodontal regenerative medicine.

We previously developed a technique whereby autogenous human mesenchymal stem cells (hMSCs) from the patient's bone marrow are combined with platelet-rich plasma for use as an alternative to such materials with predictable good prognosis (Yamada et al., 2004, 2013). However, clinical use of stem cells requires highly qualified safety investigation and quality management of cell handling, and is very expensive. These limitations currently impede the widespread use of stem cells for alveolar bone regeneration therapy. Moreover, recent studies have revealed that the implanted cells do not survive long (Ide et al., 2010; Perri et al., 2007; Toma et al., 2009). As an alternative, the effects of the secretomes, the various factors secreted into the medium, from stem cells on tissue repair and regeneration have attracted much attention (Baglio et al., 2012; Chen et al., 2008; Ciapetti et al., 2012).

We have reported the effects of the secretomes in the conditioned medium from bone marrow-derived mesenchymal stem cells (MSC-CM) on bone and periodontal tissue regeneration *in vitro* and *in vivo*. MSC-CM contains several cytokines such as insulin-like growth factor (IGF)-1, vascular endothelial growth factor (VEGF), and transforming growth factor (TGF)-β1. MSC-CM enhances cell proliferation, mobilization, angiogenesis, and expression of osteogenic markers such as *alkaline phosphatase, collagen type I,* and *Runx2* genes (Katagiri et al., 2013). MSC-CM also recruits endogenous stem cells to the grafted site and shows early bone and periodontal regeneration in rat calvarial bone defects and dog periodontal bone defects (Inukai et al., 2013; Osugi et al., 2012). Furthermore, the concentrations of cytokines contained in MSC-CM are relatively low such that use of MSC-CM does not induce the severe histological inflammatory responses that are observed with the clinical use of recombinant human BMP-2 (Katagiri et al., 2013).

In this section, our *in vitro* and *in vivo* studies about the effects of MSC-CM on bone and periodontal tissue regeneration were summarized and introduced.

2. Biological effects of MSC-CM in bore and periodontal tissue regeneration

2.1. Preparation of MSC-CM

The hMSCs were purchased from Lonza and cultured in MSC basal medium (Lonza) with MSC growth medium (SingleQuots, Lonza). Rat MSCs (rMSCs) were isolated from the femora of 7-week-old Wistar/ST rats (Japan SLC) and expanded and cultured in Dulbecco's modified Eagle medium (DMEM) (Gibco) with 10% fetal bovine serum (FBS). Cells of the second to fourth passages were used in the experiments in this study. Cells were maintained at 37°C in 5% carbon dioxide/95% air. The hMSCs were cultured in a culture dish (100 × 20 mm). When hMSCs reached 70% to 80% confluence, the medium was refreshed with 10 mL of serum-free DMEM containing antibiotics (100 units/mL penicillin G, 100 μg/mL streptomycin, and 0.25μg/mL amphotericinB; Gibco). The culture media (CM) was collected after 48 hours of incubation. This media was defined as hMSC culture-conditioned media (MSC-CM) and stored at 4°C or –80°C before being used for subsequent experiments.

Factors	Concentration (pg/mL)
IGF-1	1515.6±211.83
VEGF	465.84±108.81
TGF-β1	339.82±14.41
HGF	20.32±7.89
PDGF-BB	N.D
BMP-2	N.D
FGF-2	N.D
SDF-1	N.D

Table 1. The Levels of Cytokines Present in MSC-CM

2.2. Cytokines present in MSC-CM

The concentrations of the cytokines IGF-1, VEGF, TGF-β1, HGF, FGF-2, PDGF-BB, BMP-2, and SDF-1α in MSC-CM were quantified with ELISA. Cytokines were not detected in DMEM-0% and DMEM-30%. However, MSC-CM contained IGF-1, VEGF, TGF-β1, and HGF at concentrations of 1386 ± 465, 468.5 ± 109, 339.8 ± 14.4, and 20.3 ± 7.8 pg/ml, respectively. The other factors assayed were not detected in MSC-CM (Table 1).

2.3. MSC-CM enhances migration and proliferation of MSCs and PDLCs

The migratory properties of rMSCs and rPDLCs were examined using the CytoSelect Wound Healing Assay kit (Cell Biolabs, San Diego, CA, USA) according to the manufacturer's instructions. Briefly, cell suspension was added to the well with a plastic insert in place. The insert was removed from the well after a monolayer of cells had formed, creating a wound gap of 0.9 mm. After washing, cells were incubated at 37 °C for 48 h in MSC-CM with 30% FBS or serum-free DMEM. The extent of wound closure was determined with a light microscope (CK40; Olympus, Tokyo, Japan) at ×40 magnification. The percentage of rMSCs in the wound area of DMEM (−) was 9.28 ± 4.41. There were 70.9 ± 6.8% rMSCs in the wound area of positive control (30% FBS). MSC-CM exerted significant effects ($p < 0.05$) and closed the wound to 43.4 ± 10.6% rMSCs (Fig. 1a).

The percentage of rPDLCs in the wound area of DMEM (−) was 2.36 ± 2.32%, with 48.01 ± 6.28% in 30% FBS and 17.98 ± 4.14% in MSC-CM. Thus, MSC-CM increased rMSC migration more than four-fold and rPDLC migration more than seven-fold compared with that in DMEM (−). These differences were statistically significant ($p < 0.05$), indicating that MSC-CM enhanced rMSC and rPDLC migration and proliferation (Fig. 1b).

The level of cellular fill within the wound area in response to MSC-CM was compared with the wound-fill response in the presence of 30% FBS or serum-free DMEM as control after 48 h (a). The migration of rMSCs and rPDLSCs cultured in MSC-CM was enhanced compared with

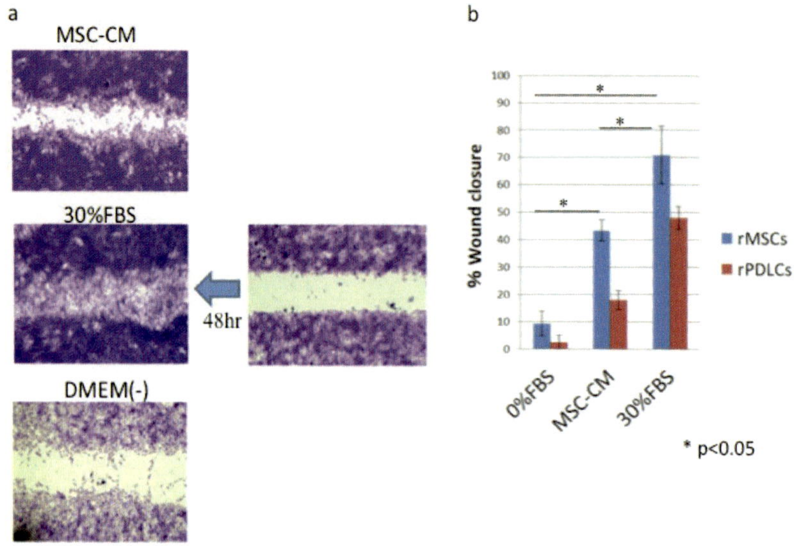

Figure 1. MSC-CM promoted migration and proliferation of rMSCs and rPDLCs in the wound-healing assay.

rMSCs and rPDLSCs cultured in DMEM (–). MSC-CM increased wound area more than 4-fold compared to that in DMEM(-) (b, *$p < 0.05$).

2.4. MSC-CM enhanced MSC migration in rat calvarial bone defect model

rMSCs were harvested and cultured as described previously and were labeled with the lipophilic tracer 1,1-dioctadecyl-3,3,3,3-tetramethylindotricarbocyanine iodide (DiR; Molecular Probes, Eugene, OR) for all imaging ex-periments (Kalchenko et al., 2006). This fluorophore is excited at 750 nm and has an emission peak at 782 nm. The cells were incubated with DiR (3×10^6 cells in 10 mL PBS containing 3.5 mg/mL dye and 0.5% ethanol) for 30 min at 37°C. The cells were then washed twice with PBS and injected intravenously into the caudal vein of Wistar/ST rats with the bone defect de-scribed previously, which had been implanted with vari-ous implant materials just before the injection of rMSCs. The rats were anesthetized by intraperitoneal injection of Somnopentyl® prior to injection. Xenogen's IVIS® 200 Series Imaging System (Xenogen, Alameda, CA) was used to monitor DiR-labeled rMSC localization within live, as well as sacrificed, animals. Imaging was performed at 1, 24, and 48h, and at 1 week after injection of DiR-labeled cells. Migration of rMSCs to the implants in vivo was analyzed in rats of the different implantation groups in which DiR-labeled rMSCs were injected into the caudal vein. Although it is not possible to detect DiR-labeled cells at a great depth with the imaging system used, this system can detect DiR-labeled cells that accumulate on the calvarial bone. The fluorescent signal in all groups increased immediately after injection. In the control Defect group, the fluorescent signal of the labeled rMSCs was only detected in the tail and abdominal region at 1, 24, and 48 h, and at 1 week after injection. We confirmed that there were no signals at the defect area in the cranium at any time point. In the PBS group, fluorescent signals were observed in the tail and in the abdominal area at 24 and 48h after injection, and very low signals were observed in the breast and cranial area after 1 week. In the MSC-CM group, a moderate increase in signal intensity was observed in the area of the tail and the abdominal region during the first 24h after injection. At 48h after injection, signal intensity in the MSC-CM-implanted area of the parietal bone started to increase. The maximum fluorescent signal in the implanted area was observed 1 week after injection (Fig. 2).

In vivo imaging analysis shows that DiR-labeled rMSCs that were injected into the caudal vein just after implantation of the materials into the calvarial bone defects started to mi-grate immediately after injection. At 24 h, 48 h, 72h and 1 week after injection, the signal of the fluorescent-labeled rMSCs was only detected in the tail and abdominal region in the control PBS group. In the MSC-CM group, a moderate increase in signal intensity was observed in the abdominal region during the first 24 h after injection. Forty-eight hours after injection, the MSC-CM-implanted area of calvarial bone started to increase in signal intensity, and, 1 week after injection, the MSC-CM implanted area, as well as the implanted bone cavity, showed the highest fluorescent signal of the experimental groups.

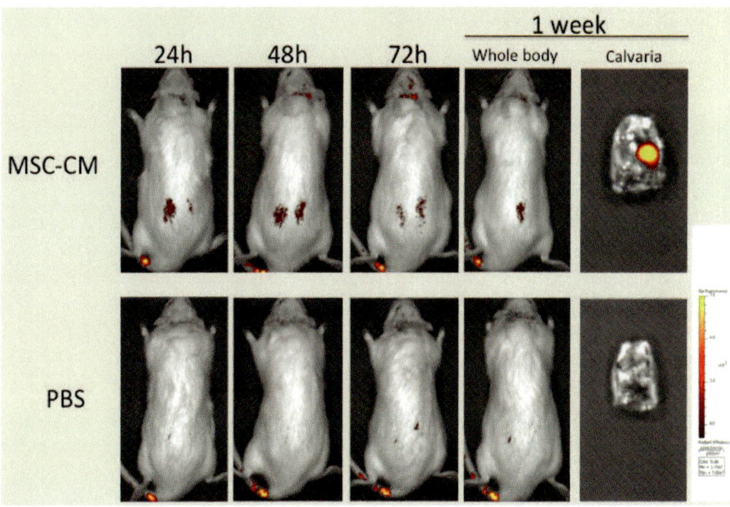

Figure 2. In vivo imaging of injected rMSC migration to implants.

2.5. MSC-CM enhanced osteogenic and angiogenic marker gene expression

The levels of expression of the *ALP*, *OCN*, *Runx2*, *VEGF-A*, *ANG-1*, and *ANG-2* genes were significantly upregulated in rMSCs cultured with MSC-CM compared with rMSCs cultured in EM (Fig. 3).

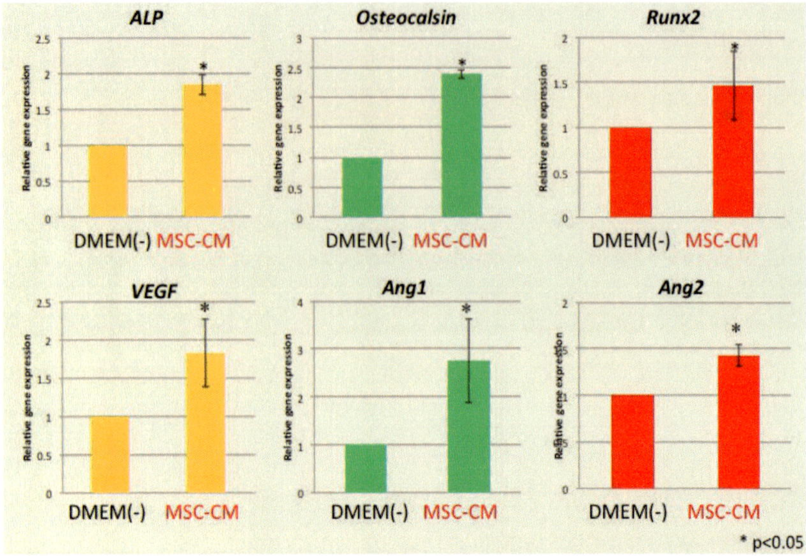

Figure 3. MSC-CM enhanced osteogenic and angiogenic marker gene expression.

The mRNA expression levels of *ALP*, *OCN*, *Runx2*, *VEGF-A*, *ANG-1*, and *ANG-2* were determined relative to the level of *Glyceraldehyde 3-phosphate dehydrogenase* (*GAPDH*) mRNA in each sample and were quantified and standardized. The level of expression of the each gene was significantly upregulated in rMSCs cultured with MSC-CM compared with rMSCs cultured in DMEM (*$p < 0.05$).

2.6. MSC-CM enhanced bone regeneration, migration of endogeneous MSCs and angiogenesis in rat calvarial bone defect model

Twenty-four 10-week-old male Wistar/ST rats were anesthetized by intraperitoneal injection of pento-barbital (Somnopentyl, Kyoritsu Seiyaku) (20 mg/kg body weight). Two circular full-thickness (through-and-through) bone defects (5 mm in diameter) were made in the calvarial bone using a trephine bur and were irrigated with saline to remove the bone debris. The experimental materials were then implanted into the defects. Atelocollagen (Terudermis, Olympus Terumo Biomaterials), which was cut into the desired form, was suspended in MSC-CM or PBS. Three groups were de-fined: *(1)* MSC-CM group: MSC-CM/Terudermis; *(2)* PBS group: PBS/Terudermis; and *(3)* defect group: unfilled defect.

Rats were sacrificed at 2 or 4 weeks after transplantation (n=4 per group). The surgical sites were dissected, fixed in 10% formalin, and subjected to micro-computed tomography (micro-CT) analysis using a laboratory x-ray CT device (LATheta, Hitachi Aloka). Images were compiled and analyzed to render three-dimensional images using OsiriX imaging software (version 3.9). Then the area (mm^2) of newly regenerated bone was compared between groups.

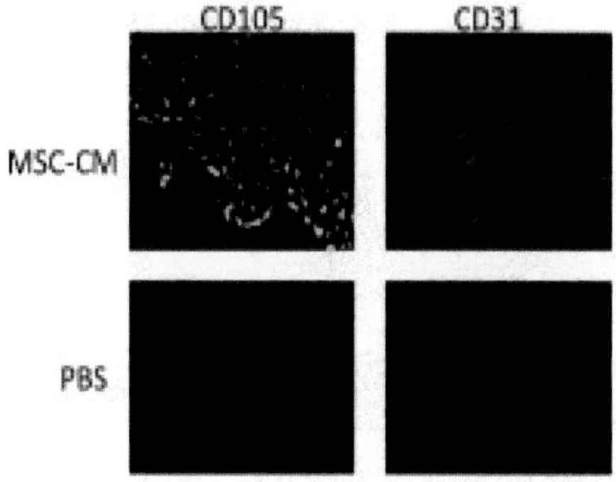

Figure 6. Immunohistochemical analysis of newly generated bone area 2 weeks after implantation.

2.7. MSC–CM enhanced periodontal tissue regeneration in dog periodontal defect model

All animal experiments were approved by the Nagoya University animal experiment committee. After a period of acclimatization of 30 days, five hybrid dogs were operated on under general anesthesia by intravenous injection of pentobarbital (Somnopentyl®) (20mg/kg body weight), and under local anesthesia with 2% lidocaine (with 1:80,000 epinephrine, ORA® Inj. Dental Cartridge; Showa Yakuhin Kako, Tokyo, Japan). Before the experimental surgery, the mandibular first and third or fourth premolars were extracted, and the extraction sites were allowed to heal for 8 weeks. For the experimental surgery, buccal and lingual mucoperiosteal flaps were elevated, and critical-size, box-type, one-wall intrabony defects (width, 4mm; height, 5 mm) were created at the distal aspect of the second, and the mesial aspect of the fourth premolars in the right and left jaw quadrants (Kim, 2005). Following root planing to remove the root cementum, a reference notch indicating a 5-mm distance from the cement-enamel junction to the bottom of the defect was made with a burr into the root surface at the base of the defects. With no differences in bone regeneration in the various grafted areas in terms of bone healing, two defects were created and implanted with two materials at random sites. An absorbable atelo-collagen sponge (TERUPLUG®; OLYMPUS TERUMO BIOMATERIALS, Tokyo, Japan) was used as a scaffold and contained 300μl MSC-CM or PBS. The dogs with defects were randomly divided into three groups (n=6 each) and implanted with graft materials: MSC-CM plus scaffold, PBS plus scaffold, or unfilled defect. The mucogingival flaps were advanced, adapted, and completely closed. Post-surgical management involved antibiotics (Azithromycin, 250 mg; Pfizer, Tokyo, Japan) daily for 3 days, a soft diet, and topical application of 2% chlorhexidine (Hibitane concentrate; Dainippon Sumitomo Pharma, Osaka, Japan) twice a week. After 4 weeks, the dogs were given general anesthesia and sacrificed by exsanguination after injection of heparin sodium (400 U/kg).

Standardized radiographic images of the defect sites were obtained with an X-ray apparatus (Dent navi Hands; Yoshida Co., Ltd., Tokyo, Japan) and dental X-ray films (BW-100; Hanshin Technical Laboratory, Nishinomiya, Japan) immediately, and 4 weeks after, transplantation.

Dental X-ray films were placed parallel to the tooth axis, and radiographic images of the defect site were ta-ken in the buccolingual direction. The defect sites were dissected and fixed in 10% neutral-buffered formalin (Wako, Japan) 4 weeks after transplantation. The specimens were decalcified in Plank-Rychro solution (Wako) for 8 weeks, routinely processed into 5μm-thick paraffin-embedded sections, stained with hematoxylin and eosin, and observed under a light microscope (Olympus). Histometric parameters were quantified using a computer-based image analysis system (Image J 1.44; National Institutes of Health). The following parameters were analyzed:

1. Cementum regeneration height: distance from the root surface notch to the coronal

extension of newly formed cementum on the root surface.

2. Bone regeneration height: distance from the root surface notch to the coronal

extension of newly formed bone along the root surface.

3. Bone regeneration area: area of new alveolar bone formed coronally from the apical

extension of the root surface notch.

Clinical healing was generally uneventful. The results from the histometric analysis are shown in Fig. 8. The cementum regeneration height, the bone regeneration height, and the bone regeneration area of the MSC-CM group were 3.01 ± 0.16 mm, 3.19 ± 0.51 mm, and 4.89 ± 1.08 mm^2, respectively. A large amount of new lamellar and woven bone formation was observed in the MSC-CM group. Thick-layered and cellular cementum on the root surface was also frequently observed in the MSC-CM group.

On the other hand, less newly regenerated bone and cementum compared to the MSC-CM group was observed in the PBS group. Dense collagen fibers were observed frequently in the PBS group. Newly regenerated bone and cementum were not apparent in the defect group. Furthermore, there was minimal inflammatory cell infiltration in the MSC-CM group compared to the other groups.

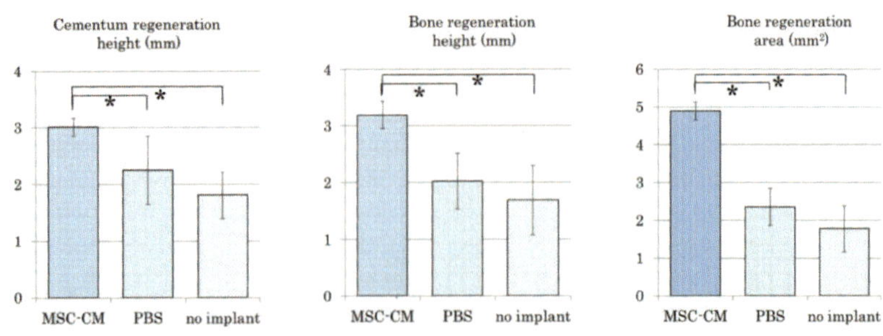

Figure 7. Representative photomicrographs and radiographic images from sites receiving experimental materials.

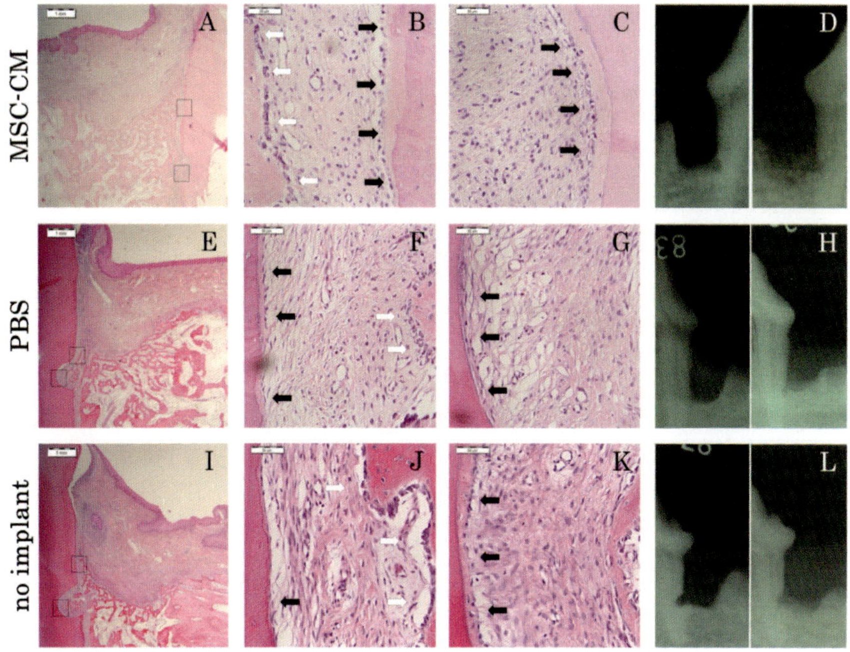

White arrows: osteoblasts. Black arrows: newly regenerated cellular cementum. Photomicrographs: hematoxylin and eosin staining.

Figure 8. (A, E and I) overview of periodontal defect site. Original magnification x12.5, scale bar=1 mm. (B, C, F, G, J and K) Higher magnification of the boxed areas. Original magnification x100, scale bar=50 lm. (D, H and L) Radiographic images were taken at 0 (left) and 4 weeks (right) after surgery. Histometric analysis of periodontal regeneration following surgical implantation of MSC-CM/TERUPLUG˙ in dog one-wall intrabony defects (means ± SD in mm or mm^2) ($p < 0.05$).

3. Conclusion

From the results of these studies, it was suggested that MSC-CM contributes to upregulation of several processes of bone and periodontal tissue regeneration through the angiogenesis and mobilization of endogenous MSCs, and thus enhanced bone and periodontal tissue regeneration. Using MSC-CM for bone and periodontal tissue regeneration may be effective because several cytokines, including MSC-CM, contribute several processes to the complex system of bone and periodontal tissue regeneration. If the MSC-CM treatment protocol is to be established for bone and periodontal tissue regeneration, it is essential that effective and therapeutic doses of MSC-CM as well as the safety of the therapy should be carefully established. Further investigation regarding these matters is now in progress.

Author details

Wataru Katagiri

Nagoya University Graduate School of Medicine, Japan

References

[1] Arrington ED, Smith WJ, Chambers HG, Bucknell AL & Davino NA. (1996). Complications of iliac crest bone graft harvesting. *Clin Orthop Relat Res,* Vol.329, (August 1996), pp.300-309, ISSN 0009-921X.

[2] Baglio SR, Pegtel DM & Baldini N. (2012). Mesenchymal stem cell secreted vesicles provide novel opportunities in (stem) cell-free therapy. *Front Physiol* Vol 3, (September 2012), pp.1-11, ISSN 1664-042X.

[3] Chen L, Tredget EE, Wu PYG & Wu Y. (2008). Paracrine factors of mesenchymal stem cells recruit macrophages and endothelial lineage cells and enhance wound healing. *PloS ONE,* Vol. 3, No.4, (June 2008), pp.e1886, ISSN 1932-6203.

[4] Ciapetti G, Granchi D & Baldini N. (2012). The combined use of mesenchymal stromal cells and scaffolds for bone repair. *Curr Pharm Des, Vol.* 18, No.13, (July 2012), pp. 1796–1820, ISSN 1873-4276.

[5] Damien CJ & Parsons JR. (1991). Bone graft and bone graft substitutes: a review of current technology and applications. *J Appl Biomater* Vol. 2, No.3, (June 1996), pp. 187-208, ISSN 1045-4861.

[6] Herford AS & Boyne PJ. (2008), Reconstruction of mandibular continuity defects with bone morphogenic protein-2 (rhBMP-2). *J Oral Maxillofac Surg,* Vol.66, No.4, (April 2008), pp.616-624, ISSN 1531-5053.

[7] Ide C, Nakai Y, Nakano N, Seo T, Yamada Y, Endo K, Noda T, Saito F, Suzuki Y, Fukushima M & Nakatani T. (2010). Bone marrow stromal cell transplantation for treatment of sub-acute spinal cord injury in rat. *Brain Res,* Vol.1332, (May 2010), pp.32-47, ISSN 1872-6240.

[8] Inukai T, Katagiri W, Yoshimi R, Osugi M, Kawai T, Hibi H & Ueda M. (2013). Novel application of stem cell-derived factors for periodontal regeneration. *Biochem Biophys Res Commun,* Vol.430, No. 2, (January 2013), pp.763-768, ISSN 1090-2104.

[9] Joshi A & Kostakis GC. (2004). An investigation of post-operative morbidity following iliac crest graft harvesting. *Br Dent J,* Vol.196, No. 3, (March 2004), pp.167-171, ISSN 0007-0610.

[10] Kalchenko V, Shivtiel S, Malina V, Lapid K, Haramati S, Lapidot T, Brill A, & Harmelin A. (2006). Use of lipophilic near-infrared dye in whole-body optical imaging of

hema-topoietic cell homing. *J Biomed Opt Vol.* 11, No.5, (January 2007), pp.50507, ISSN 1083-3668.

[11] Katagiri W, Osugi M, Kawai T & Ueda M. (2013). Novel cell-free regenerative medicine of bone using stem cell derived factors. *Int J Oral Maxillofac Implants* Vol. 28, No. 4, (March 2014), pp.1009-1016, ISSN 1942-4434.

[12] Kawasaki K, Aihara M, Honmo J, Sakurai S, Fujimaki Y, Sakamoto K, Fujimaki E, Wozney JM & Yamaguchi A. (1998). Effects of reconmbinant human bone morphogenetic protein-2 on differentiation of cells isolated from human bone, muscle, and skin. Bone 1998: 23, No.3, (December 1998), pp223-231, ISSN 8756-3282.

[13] Kim CS, Choi SH, Cho KS, Chai JK, Wikesjö UM & Kim CK. (2005). Periodontal healing in one-wall intra-bony defects in dogs following implantation of autogenous bone or a coral-derived biomaterial. *J Clin Periodontol*, Vol.32, No.6, (January 2005), pp.583–589, ISSN 0303-6979.

[14] Langer R & Vacanti JP. (1993). Tissue engineering. *Science*, Vol.260, No.5110, (June 1993), pp.920-926, ISSN 0036-8075.

[15] Osugi M, Katagiri W, Yoshimi R, Inukai T, Hibi H & Ueda M. (2012). Conditioned media from mesenchymal stem cells enhancedbone regeneration in rat calvarial bone defects. *Tissue Eng Part A*, Vol. 18, No. 13-14, (July 2012), pp.14779-1489, ISSN 1937-335X.

[16] Perri B, Cooper M, Lauryssen C & Anand N. (2007). Adverse swelling associated with use of rh-BMP-2 in anterior cervical discectomy and fusion: a case study. *Spine J*, Vol. 7, No.2, (May 2007), pp.235-239, ISSN 1529-9430.

[17] Toma C, Wagner WR, Bowry S, Schwartz A & Villanueva F. (2009). Fate of culture-expanded mesenchymal stem cells in the microvasculature: in vivo observations of cell kinetics. *Circ Res*, Vol.104, No.3, (February 2009), pp.398-402, ISSN 1524-4571.

[18] Vaidya R, Carp J, Sethi A, Bartol S, Craig J & Les CM. (2007). Complications of anterior cervical discectomy and fusion using recombinant human bone morphogenetic protein-2. *Eur Spine J*, Vol.16, No.8, (November 2007), pp.1257-1265, ISSN 0940-6719.

[19] Yamada Y, Ueda M, Naiki T, Takahashi M, Hata K & Nagasaka T. (2004). Autogenous injectable bone for regeneration with mesenchymal stem cells (MSCs) and platelet-rich plasma (PRP) – Tissue-engineered bone regeneration. T*issue Eng*, Vol.10, No.5-6, (January 2005), pp.955-964, ISSN 1076-3279.

[20] Yamada Y, Nakamura S, Ito K, Umemura E, Hara K, Nagasaka T, Abe A, Baba S, Furuichi Y, Izumi Y, Klein OD & Wakabayashi T. (2013). Injectable bonetissue engineeringusingexpanded mesenchymal stem cells. *Stem Cells*, Vol.31, No.3, pp.572-580, ISSN 1549-4981.

Chapter 9

Whole Tooth Regeneration Using a Bioengineered Tooth

Masamitsu Oshima and Takashi Tsuji

Additional information is available at the end of the chapter

http://dx.doi.org/10.5772/58908

1. Introduction

The tooth, which is an ectodermal organ whose development is regulated by reciprocal epithelial-mesenchymal interactions (Jussila et al., 2013), contributes to oral functions associated with mastication and enunciation, which are important aspects of general health and quality of life (Proffit et al., 2004). Teeth have a three-dimensional multicellular structure composed of characteristic hard tissues, *e.g.*, enamel, dentin, cementum and alveolar bone. Teeth also have soft connective tissues, such as pulp and periodontal ligaments, which contain nerve fibres and blood vessels that are important for maintaining tooth homeostasis (Avery, 2002). Dental caries, periodontal disease and trauma, which have high prevalence rates in dental disorders, cause fundamental problems for oral function and are associated with oral and general health issues (Proffit et al., 2004). To restore occlusal function after tooth loss, conventional dental treatments based on replacing teeth with artificial materials, such as fixed or removable dentures, have been established. Dental implants, which are able to stand alone in the jawbone without invading the adjacent teeth, have been used for the rehabilitation of tooth loss. Although these artificial therapies are widely applied to treat dental disorders, recent advances in tissue regeneration have been made that enhance the functions of the biological tooth, allow for underlying tooth movement through bone remodelling and aid the ability to perceive noxious stimuli (Huang et al., 2009). Substantial advances in the development of regenerative therapies have been driven by our understanding of embryonic development, stem cell biology and tissue engineering technologies (Yelick & Vacanti, 2006). Currently, an important concept in regenerative therapy is the transplantation of tissue-derived stem cells or *in vitro*-manipulated induced pluripotent stem (iPS) cells (Volponi et al., 2010). These therapies are attractive therapeutic concepts that have the potential to repair damaged tissues and restore the partial loss of organ function (Korbling & Estrov, 2003). In dental medicine, tooth tissue-derived stem cells and the cytokine network that regulates tooth development have been well characterised at the molecular level (Jussila et al., 2013). These advances can be applied to the repair of dental pulp and periodontal tissues, including the alveolar bone (Egusa et al., 2012, 2013). Organ replacement regenerative therapy, which involves constructing a fully functional bioengineered organ using three-dimensional cell manipulation *in vitro*, holds great promise for the replacement of dysfunctional organs following disease, injury or aging. Tooth regenerative therapy would also involve the replacement of a lost or damaged tooth with a bioengineered tooth, constructed with stem cells, that has the capacity to become a functional unit comprising the whole tooth and periodontal tissue (Yen & Sharpe, 2008). It is anticipated that tooth replacement therapy will be established in the near future as a novel biological treatment for the functional recovery of lost teeth to satisfy both aesthetic and physiological requirements (Fig. 1). Over the past three decades, many approaches for replacing missing teeth have been studied, including three-dimensional bioengineered teeth and tooth germ generation using biodegradable materials and cell aggregation methods (Volponi et al., 2010). Recently, studies have reported tooth replacement by transplantation of fully functioning bioengineered teeth having the correct tooth structure, masticatory performance, proper responsiveness to mechanical stress and neural function after transplantation into the region of tooth loss (Ikeda et al., 2009; Nakao et al., 2007; Oshima et al., 2011). In this chapter, we describe novel technologies for whole tooth replacement that have the potential to provide functional recovery and could someday replace current dental treatments based on artificial materials.

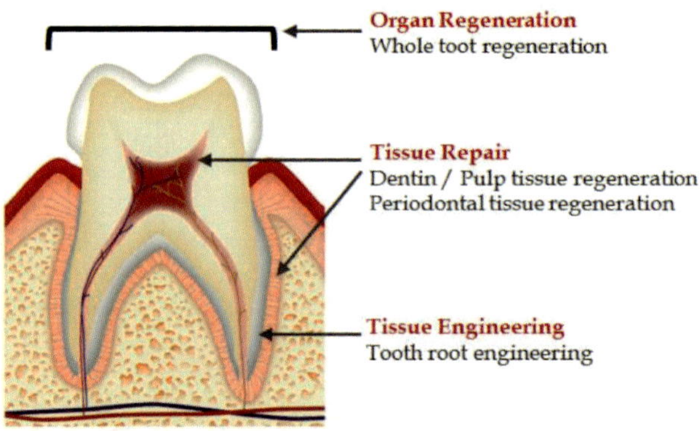

Figure 1. Concepts of tooth regenerative therapy

Recent approaches for tooth regenerative therapy have included tissue repair and whole tooth replacement. Tooth regenerative therapy and stem cell transplantation therapies are regarded as attractive approaches for repairing tissue that has been damaged by dental caries or periodontal disease. The transplantation of dental stem cells has been examined for the treatment of dental caries, pulp injury and periodontal disease.

2. The mechanisms of tooth development

Ectodermal organs, such as the teeth, hair and salivary glands, arise from their respective organ germs through reciprocal epithelial-mesenchymal interactions in the developing embryo. These interactions, which involve various signalling molecules and transcription factors, are the principal mechanism regulating organogenesis (Jussila et al., 2013). In tooth germ development, the dental lamina first thickens (lamina stage). This stage is followed by epithelial thickening (placode stage) at the future location of the tooth and subsequent epithelial budding to the underlying neural crest-derived ecto-mesenchyme. Tooth germ formation is initiated on embryonic days (EDs) 10-11 in mice by epithelial signals that include fibroblast growth factor (FGF) 8, bone morphogenetic protein (BMP) 4, sonic hedgehog (Shh), tumour necrosis factor (TNF) and Wnt10b. These signals induce the expression of several transcription factors in the dental mesenchyme that condense around the developing epithelial bud (bud stage) (Jussila et al., 2013; O'connell et al., 2013). At ED13.5-14.5, the first enamel knot, which acts as a signalling centre to orchestrate tooth development by controlling the gene expression of various signalling molecules and transcription factors, is formed in the dental epithelium (cap stage). At ED16-18, the epithelial and mesenchymal cells in the tooth germ terminally differentiate into the tooth-tissue progenitor cells, such as ameloblasts, odontoblasts, and dental follicle cells (bell stage). Ameloblasts and odontoblasts accumulate the enamel and dentin matrix, respectively, at the boundary surface between the epithelium and mesenchyme, while dental follicle cells differentiate into the periodontal tissues, which include the cementum, periodontal ligaments and alveolar bone (Avery, 2002).

3. A novel three-dimensional cell manipulation method for whole tooth regeneration

One current biological approach for the regeneration of three-dimensional organs is based on recapitulating organogenesis by mimicking the reciprocal epithelial-mesenchymal interactions that occur in the developing embryo, thereby developing fully functional bioengineered organs from a bioengineered organ germ generated from immature stem cells via three-dimensional cell manipulation *in vitro*. For tooth regeneration, one proposed concept has been to transplant a bioengineered tooth germ into the recipient jaw and allowing it to develop into a functional mature tooth *in situ* (Fig. 2, upper). It is also expected that it will be possible to transplant a bioengineered tooth unit that includes mature tooth, periodontal ligament and alveolar bone, which will achieve biological engraftment through bone integration with the recipient's jaw (Fig. 2, lower).

To realise whole tooth replacement, the first critical issue is to develop a three-dimensional cell manipulation method using completely dissociated epithelial and mesenchymal cells *in vitro*. Previously, it has been reported that using a polyglycolic acid and poly-L-lactate-co-glycolide copolymer (PLA/PLGA) or a collagen sponge as a tooth-shaped scaffold and seeding them with epithelial and mesenchymal cells isolated from tooth buds could generate small tooth structures (Honda et al., 2007; Yelick & Vacanti, 2006).

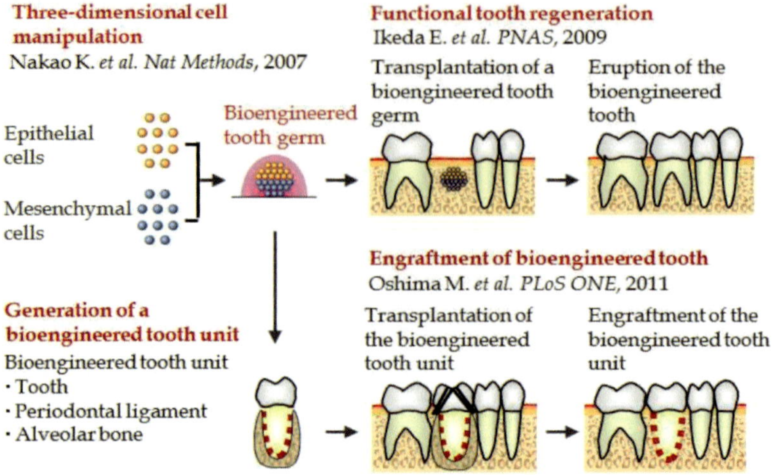

Figure 2. Strategies for whole tooth replacement via regenerative therapies. Functioning teeth can now be regenerated *in vivo* by transplanting bioengineered tooth germ generated from epithelial and mesenchymal cells via the organ germ method, or bioengineered tooth units with periodontal ligament and alveolar bone developed from bioengineered tooth germ.

In addition, the cell aggregation method, which aims to reconstitute a bioengineered organ germ, has been applied for the transplantation of cell aggregates constructed from dental epithelial and mesenchymal cells, and it has been reported that this approach can generate appropriate tooth formation (Hu et al., 2006). It has also been reported that mixed cell aggregates of tooth germ-derived epithelial and mesenchymal cells can develop into a tooth with the correct structure, following epithelial cell sorting and subsequent self-organisation of the epithelial and mesenchymal cells (Song et al., 2006). However, these approaches suffer from critical limitations, including a low frequency of tooth formation and irregularity of the resulting tooth tissue structures, for example with enamel-dentin complex formation and the arrangements of the ameloblast/odontoblast cell lineages.

To achieve precise replication of the processes in organogenesis, an *in vitro* three-dimensional novel cell manipulation method designated as the bioengineered organ germ method has been developed (Nakao et al., 2007). This innovative method is based on compartmentalisation of the epithelial and mesenchymal cells at a high-cell density in a type I collagen gel (Fig. 3A). Bioengineered tooth germ created by this technique, which could allow for large-scale organ development, mimics the multicellular assembly underlying epithelial-mesenchymal interactions during natural tooth development. This bioengineered tooth germ generates a correct tooth structure after transplantation in an organ culture *in vitro* as well as following placement into a subrenal capsule *in vivo*. The bioengineered tooth germ generated by this method was also found to develop in the oral cavity to form the proper tooth structure (Nakao et al., 2007). Furthermore, this unique technology can successfully generate a size-controlled bioengineered tooth unit comprising a mature tooth, periodontal ligament and alveolar bone after transplantation into the subrenal capsule (Fig. 3B). These technologies have the potential to be adapted for successful functional tooth replacement *in vivo* and are expected to represent a substantial advance in bioengineered organ replacement regenerative therapy.

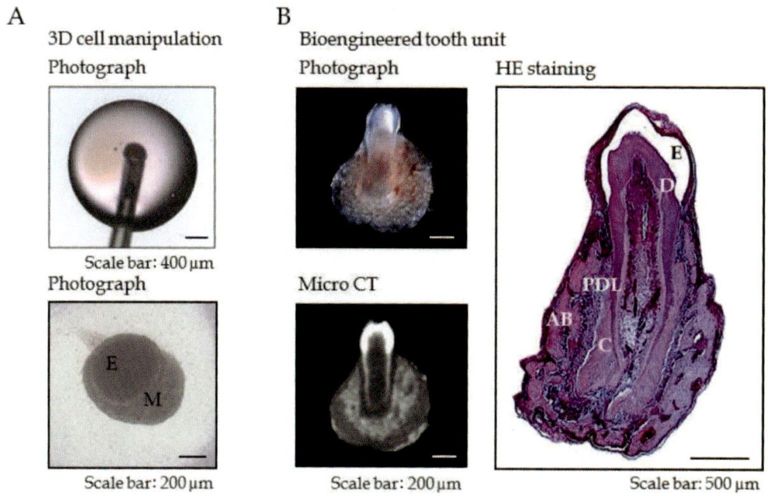

Figure 3. The organ germ method: three-dimensional cell processing A) Dissociated mesenchymal cells at a high density are injected into the centre of a collagen drop. Dissociated tooth germ-derived epithelial cells are subsequently injected into the drop adjacent to the mesenchymal cell aggregate (upper). Within 1 day of organ culture, bioengineered tooth germ formation with appropriate compartmentalisation between epithelial and mesenchymal cells and cell-to-cell compaction was observed (lower). B) By transplanting a bioengineered tooth germ into a subrenal capsule for 30 days (left panel), a bioengineered tooth unit comprising a mature tooth with the correct structural components such as enamel (E), dentin (D), periodontal ligament (PDL) and alveolar bone (AB) can be produced (right panel).

4. Functional tooth replacement therapy

Oral functions such as mastication, pronunciation, and facial aesthetics have an important influence on quality of life because they facilitate both oral communication and nutritional intake. These functions are achieved with the teeth, masticatory muscles and the temporomandibular joint, under control of the central nervous system. For the realisation of tooth replacement regenerative therapy, a regenerated tooth developing from bioengineered germ tissue or a transplanted bioengineered mature tooth unit must be capable of properly engrafting into the lost tooth region in an adult oral environment and acquiring full functionality, including sufficient masticatory performance, biochemical cooperation with periodontal tissues and afferent responsiveness to noxious stimulations in the maxillofacial region (Proffit et al., 2004).

4.1. Transplantation of bioengineered tooth germ or a bioengineered mature tooth unit as a tooth replacement therapy

The critical issue dictating the success of tooth regenerative therapy via the transplantation of bioengineered tooth germ tissue into the lost tooth region is whether the germ can erupt and occlude properly with the opposing tooth in an adult oral environment. It has previously been demonstrated that transplanted natural tooth germ erupts in a murine toothless diastema region (Ohazama et al., 2004). We have also reported that a bioengineered tooth germ can develop the correct tooth structure in an oral cavity and successfully erupt 37 days after transplantation (Ikeda et al., 2009). The bioengineered tooth subsequently reached the occlusal plane and achieved occlusion with the opposing tooth from 49 days onwards (Fig. 4A, B). In the case of a transplanted bioengineered mature tooth unit comprising mature tooth, periodontal ligament and alveolar bone, the most critical consideration is whether that unit can be engrafted into the tooth loss region through bone integration, which involves natural bone remodelling in the recipient. A bioengineered tooth unit transplanted at a position reaching the occlusal plane with the opposing upper first molar was successfully engrafted after 40 days and thereafter maintained the periodontal ligament originating from the bioengineered tooth unit through successful bone integration (Fig. 4C) (Oshima et al., 2011). The enamel and dentin hardness of the bioengineered tooth components were in the normal range when analysed by the Knoop hardness test (Ikeda et al., 2009; Oshima et al., 2011). These approaches demonstrate the potential to successfully recover masticatory performance and natural tooth tissue through state-of-the-art bioengineering technology.

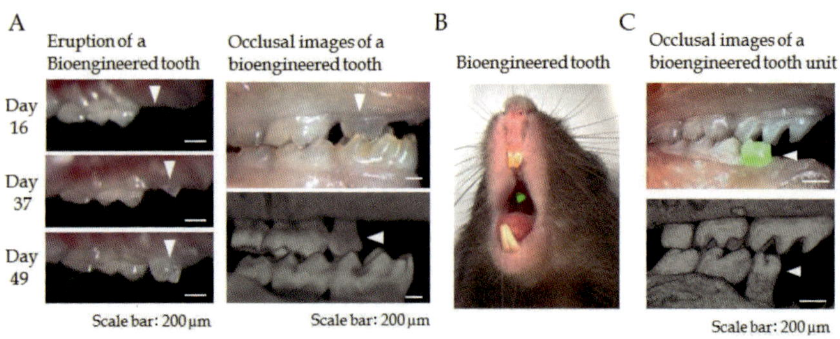

Figure 4. Regeneration of a bioengineered tooth in an adult oral environment A) A transplanted bioengineered tooth germ erupted and reached the occlusal plane with the opposing lower first molar 49 days after transplantation. B) GFP-labelled bioengineered tooth (right panel) erupted in the oral environment of adult mice. C) A bioengineered tooth unit was engrafted by bone integration and reached the occlusal plane with the opposing upper first molar at 40 days post transplantation.

4.2. Biological response of bioengineered teeth to mechanical stress

Biological oral functions require cooperation between teeth and the maxillofacial region through the connection of periodontal ligaments (Dawson, 2006). Tooth loss and periodontal disease cause fundamental problems for oral function, including mastication, as well as associated health issues. The periodontal ligament plays an essential role in the pathogenic and physiological tooth response to extreme mechanical forces from bone remodelling accompanied by orthodontic tooth movement (Proffit et al., 2004). Studies on autologous tooth transplantation have indicated that healthy periodontal tissue remaining on the tooth root can successfully restore physiological tooth function, including bone remodelling, and effectively prevent ankylosis. In contrast, the absence of a periodontal ligament in osseointegrated dental implants is associated with deficiencies in essential tooth functions and in the natural structural relationship between the tooth root and alveolar bone (Dawson, 2006). The periodontal ligament of bioengineered teeth that erupted following the transplantation of bioengineered tooth germ and mature tooth units achieved functional tooth movement comparable with that of natural teeth. Bioengineered teeth also successfully underwent bone remodelling in response to mechanical stress via the proper localisation of osteoclasts and osteoblasts, indicating that a bioengineered tooth can reproduce critical tooth functions by restoring and re-establishing cooperation with the surrounding jawbone (Ikeda et al., 2009; Oshima et al., 2011).

4.3. Perceptive neuronal potential of bioengineered teeth

The peripheral nervous system is established by the growth of axons that navigate and establish connections with developing target organs during embryogenesis (Guyton & Hall, 2000). The perceptive potential for noxious stimulation, including mechanical stress and pain, is important for proper organ function (Guyton & Hall, 2000). Additionally, it is believed that the recovery of the nervous system, which requires the re-entry of nerve fibres following organ transplantation, is critical for reconstituting organ function. Teeth are a peripheral organ for sensory and sympathetic nerves, both of which play important roles in tooth function and protection (Dawson, 2006). It is anticipated that tooth regenerative therapies will be able to recover the neuronal ability related to the perception of mechanical forces that are lacking in implant patients. Importantly, sensory and sympathetic nerve fibres innervate both the pulp and periodontal ligament of a bioengineered tooth following its eruption (Ikeda et al., 2009). Thus, these bioengineered teeth possess appropriate perceptive potential for nociceptive pain stimulations, such as pulp injury and orthodontic treatment, and can properly transduce these events to the central nervous system through c-Fos immunoreactive neurons (Ikeda et al., 2009; Oshima et al., 2011). In this way, bioengineered teeth can indeed restore the perceptive potential for noxious stimuli in cooperation with the maxillofacial region.

5. Future directions for tooth regeneration

To realise the use of tooth regenerative therapy in future clinical applications, one of the major research hurdles remaining is the identification of appropriate cell sources. The cell source may be optimised by using the patient's own cells for regenerative therapy to avoid immunological rejection. Tooth tissue-derived stem cells found in pulp and periodontal ligaments can differentiate into dental cell lineages and contribute to the supply of various progenitor cells (Egusa et al., 2012, 2013). While these tissues are good candidate cell sources for stem cell transplantation therapy for tooth tissue repair, epithelial-mesenchymal interaction driven tooth inductive potential has not been reported for these stem cells. Other candidate cell sources for whole tooth regeneration include embryonic stem (ES) cells and iPS cells, which are capable of differentiating into endoderm, ectoderm and mesoderm (Takahashi et al., 2006). Recently, iPS cells have been established from various oral tissues, and reprogramming procedures for dental epithelial and mesenchymal fates have been established (Arakaki et al., 2013; Otsu et al., 2013). Another important direction for future research on tooth regenerative therapies is the identification of key factors for reprogramming non-dental cells into dental epithelium and mesenchyme. Notably, the self-organisation of various tissues such as the optic cup and adenohypophysis using uniform pluripotent stem cells in three-dimensional culture has been reported (Eiraku et al., 2011; Suga et al., 2011). A three-dimensional *in vitro* organogenesis system using appropriately induced stem cells will be essential for the regenerative replacement of whole teeth and other organs (Sasai et al., 2013). These approaches will contribute to the realisation of future tooth replacement regenerative therapies.

6. Conclusion

The technology of regenerative medicine has progressed remarkably, and many patients and clinicians are anticipating the realisation of whole tooth regenerative therapy. Tooth regenerative therapy is now regarded as a crucial model for future organ replacement regenerative therapies for severe diseases and will contribute substantially to the understanding of tissue regeneration for more complex organs.

Acknowledgements

This work was partially supported by Health and Labour Sciences Research Grants from the Ministry of Health, Labour, and Welfare (No. 21040101) to Akira Yamaguchi (Tokyo Medical and Dental University), a Grant-in-Aid for Scientific Research (A) (No. 20249078) to T. Tsuji (2008–2010) and a Grant-in-Aid for young Scientists (B) to M. Oshima from the Ministry of Education, Culture, Sports and Technology, Japan.

http://dx.doi.org/10.5772/58908

Author details

Masamitsu Oshima[1] and Takashi Tsuji[2,3,4]

1 Department of Oral Rehabilitation and Regenerative Medicine, Graduate School of Medicine, Dentistry and Pharmaceutical Sciences, Okayama University, Okayama, Japan

2 RIKEN Center for Developmental Biology, Kobe, Hyogo, Japan

3 Research Institute for Science and Technology, Tokyo University of Science, Noda, Chiba, Japan

4 Organ Technologies Inc., Tokyo, Japan

References

[1] Arakaki, M.; Ishikawa, M.; Nakamura, T.; Iwamoto, T.; Yamada, A.; Fukumoto, E.; Saito, M.; Otsu, K.; Harada, H.; Yamada, Y. & Fukumoto, S. (2012). Role of epithelial-stem cell interactions during dental cell differentiation. *J Biol Chem.*, Vol. 287, pp. 10590-10601

[2] Avery, J. K. (2002). *Oral development and histology*, Steele, P. F. & Avery, N., (Eds.), pp. 153-212, Thieme Press, ISBN 1-58890-028-2, New York, U. S. A.

[3] Dawson, P. E. (2006). *Functional occlusion: from TMJ to smile design*, pp. 18-26, Mosby Press, ISBN 978-0-323-03371-8, St. Louis, U. S. A.

[4] Eiraku, M.; Takata, N.; Ishibashi, H.; Kawada, M.; Sakakura, E.; Okuda, S.; Sekiguchi, K.; Adachi, T. & Sasai, Y. (2011). Self-organizing optic-cup morphogenesis in three-dimensional culture, *Nature*, Vol. 472, pp. 51-56

[5] Egusa, H.; Sonoyama, W.; Nishimura, M.; Atsuta I. & Akiyama, K. (2012). Stem cells in dentistry--part I: Stem cell sources, *J Prosthodont Res.*, Vol. 56 No.3, pp.151-65

[6] Egusa, H.; Sonoyama, W.; Nishimura, M.; Atsuta I. & Akiyama, K. (2013). Stem cells in dentistry--part II: Clinical applications, *J Prosthodont Res.*, Vol. 56, No. 4, pp. 229-48

[7] Guyton, A.C. and Hall, J.E. (2000). *Textbook of medical physiology*, Saunders, ISBN 978-1416045748, U. S. A.

[8] Honda, M.J.; Tsuchiya, S.; Sumita, Y.; Sagara, H. & Ueda, M. (2007). The sequential seeding of epithelial and mesenchymal cells for tissue-engineered tooth regeneration., *Biomaterials*, Vol. 28, pp. 680-689

[9] Hu, B.; Nadiri, A.; Kuchler-Bopp, S.; Perrin-Schmitt, F.; Peters, H. & Lesot, H. (2006). Tissue engineering of tooth crown, root, and periodontium. *Tissue Eng,* Vol. 12, pp. 2069-2075

[10] Huang, G. T.; Gronthos, S. & Shi, S. (2009). Mesenchymal stem cells derived from dental tissues vs. those from other sources: their biology and role in regenerative medicine, *J Dent Res*, Vol. 88, No. 9, pp. 792-806

[11] Ikeda, E.; Morita, R.; Nakao, K.; Ishida, K.; Nakamura, T.; Takano-Yamamoto, T.; Ogawa, M.; Mizuno, M.; Kasugai, S. & Tsuji, T. (2009). Fully functional bioengineered tooth replacement as an organ replacement therapy. *Proc Natl Acad Sci U S A,* Vol. 106, No. 32, pp. 13475-13480

[12] Jussila, M.; Juuri, E. & Thesleff, I. (2013). Tooth morphogenesis and renewal. *Stem Cells in Craniofacial Development and Regeneration,* George, T.-J. Huang & Thesleff, I (Eds), pp. 109-134, Wiley-Blackwell. ISBN 978-1118279236, New jersey, U.S.A.

[13] Korbling, M. & Estrov, Z. (2003). Adult stem cells for tissue repair-a new therapeutic concept ?, *N Engl J Med*, Vol. 349, No.6, pp. 570-582

[14] Nakao, K.; Morita, R.; Saji, Y.; Ishida, K.; Tomita, Y.; Ogawa, M.; Saitoh, M.; Tomooka, Y. & Tsuji, T. (2007) The development of a bioengineered organ germ method, *Nat Methods*, Vol. 4, No. 3, pp. 227-230

[15] O'connell, D.J.; Ho, J.W.O. & Mass, R.L. (2013). System biology of early tooth development. *Stem Cells in Craniofacial Development and Regeneration,* George, T.-J. Huang & Thesleff, I (Eds), pp. 179-202, Wiley-Blackwell. ISBN 978-1118279236, New jersey, U.S.A.

[16] Ohazama, A.; Modino, S. A.; Miletich, I. & Sharpe, P. T. (2004). Stem-cell-based tissue engineering of murine teeth, *J Dent Res,* Vol. 83, No. 7, pp. 518-522

[17] Oshima, M.; Mizuno, M.; Imamura, A.; Ogawa, M.; Yasukawa, M.; Yamazaki, H.; Morita, R.; Ikeda, E.; Nakao, K.; Takano-Yamamoto, T.; Kasugai, S.; Saito, M. & Tsuji, T. (2011). Functional tooth regeneration using abioengineered tooth unit as a mature organ replacement regenerative therapy. *PLoS One,* Vol. 6, No. 7, e21531

[18] Otsu, K.; Kishigami, R.; Oikawa-Sasaki, A.; Fukumoto, S.; Yamada, A.; Fujiwara, N.; Ishizeki, K. & Harada, H. (2012). Differentiation of induced pluripotent stem cells into dental mesenchymal cells, *Stem Cells Dev.,* Vol. 21, pp. 1156-1164

[19] Proffit, W. R; Fields, H. W. Jr. & Sarver. D. M. (2004) *Contemporary orthodontics,* pp. 78-83, Mosby Press, ISBN 978-0323040464, St. Louis, U. S. A.

[20] Sasai, Y. (2013). Next-generation regenerative medicine: organogenesis from stem cells in 3D culture, *Cell Stem Cell.* Vol. 12, No. 5, pp. 520-30

[21] Song, Y.; Zhang, Z.; Yu, X.; Yan, M.; Zhang, X.; Gu, S.; Stuart, T.; Liu, C.; Reiser, J.; Zhang, Y. & Chen, Y. (2006). Application of lentivirus-mediated RNAi in studying

gene function in mammalian tooth development. *Dev Dyn,* Vol. 235, No. 5, pp. 1334-1344

[22] Suga, H.; Kadoshima, T.; Minaguchi, M.; Ohgushi, M.; Soen, M.; Nakano, T.; Takata, N.; Wataya, T.; Muguruma, K.; Miyoshi, H.; Yonemura, S.; Oiso, Y. & Sasai Y. (2011). Self-formation of functional adenohypophysis in three-dimensional culture, *Nature,* Vol. 480, pp. 57-62

[23] Takahashi, K. & Yamanaka, S. (2006). Induction of pluripotent stem cells from mouse embryonic and adult fibroblast cultures by defined factors. *Cell,* Vol. 126, pp. 663-676

[24] Volponi, A.A.; Pang, Y. & Sharpe, P.T. (2010). Stem cell-based biological tooth repair and regeneration. *Trends in cell biology.* Vol. 20, pp. 715-722

[25] Yelick, P.C. & Vacanti, J.P. (2006). Bioengineered teeth from tooth bud cells. *Dent Clin North Am.,* Vol. 50, pp. 191-203

[26] Yen, A. H. & Sharpe, P. T. (2008). Stem cells and tooth tissue engineering, *Cell Tissue Res,* Vol. 331, No. 1, pp. 359-372